Applying
Guiding Principles
of Effective
Program Delivery

Best Practices and Advances
in Program Management Series

Series Editor
Ginger Levin

PUBLISHED TITLES

Construction Program Management
Joseph Delaney

Applying Guiding Principles of Effective Program Delivery
Kerry R. Wills

Program Management: A Life Cycle Approach
Ginger Levin

*Implementing Program Management: Templates and Forms Aligned
with the Standard for Program Management,
Third Edition* (2013) *and Other Best Practices*
Ginger Levin and Allen R. Green

FORTHCOMING TITLES

Program Governance
Muhammad Ehsan Khan

*Successful Program Management:
Complexity Theory, Communication, and Leadership*
Wanda Curlee and Robert Lee Gordon

Sustainable Program Management
Gregory T. Haugan

The Essential Program Management Office
Gary Hamilton

*Leading Virtual Project Teams: Adapting Leadership Theories
and Communications Techniques to 21st Century Organizations*
Margaret R. Lee

From Projects to Programs: A Project Manager's Journey
Samir Penkar

Applying Guiding Principles of Effective Program Delivery

Kerry R. Wills

CRC Press
Taylor & Francis Group
Boca Raton London New York

CRC Press is an imprint of the
Taylor & Francis Group, an **informa** business
AN AUERBACH BOOK

CRC Press
Taylor & Francis Group
6000 Broken Sound Parkway NW, Suite 300
Boca Raton, FL 33487-2742

First issued in paperback 2019

© 2014 by Taylor & Francis Group, LLC
CRC Press is an imprint of Taylor & Francis Group, an Informa business

No claim to original U.S. Government works

ISBN-13: 978-1-4665-8789-2 (hbk)
ISBN-13: 978-0-367-37973-5 (pbk)

Visit the Taylor & Francis Web site at
http://www.taylorandfrancis.com

and the CRC Press Web site at
http://www.crcpress.com

Contents

x • *Contents*

Preface

I have been managing large technology programs and have been a student of the project and program management disciplines for nearly two decades. Based on my observations and personal experience I believe that the most successful programs are ones that follow a certain set of guiding principles and are not necessarily the programs that have the best processes or tools. The irony is that most of the materials in the marketplace focus on specific techniques or functions and not on how to apply principles to use them effectively and optimally.

As a result of not seeing many materials regarding this type of approach, I made the decision to write this book in the context of these core guiding principles of a consultative approach and how to apply them. Examples of the guiding principles are diligence, transparency, and a single source of truth for key program information. The cover of this book shows the Pyramids of Giza because they are a perfect example of a large successful program that followed the principles of diligence and attention to detail.

The book is organized into five chapters:

- The first chapter sets the context and case for the consultative approach.
- The second chapter then defines the consultative approach and the associated program management functions.
- The third chapter describes each of the guiding principles and provides a case study for each that demonstrates the use of the principle.
- The fourth chapter summarizes each of the program management functions and then shows how the guiding principles are applied to that function.
- The fifth chapter then provides a summary of the concepts and also highlights several key themes from within the book. It also provides some examples of the templates suggested in the book.

Acknowledgments

It is important to recognize those people who helped and supported me in writing this book.

- My wife, Diane, for supporting me during the many hours I was locked away in my study writing this book
- My kids, Stephanie and Matthew, for keeping me young and energized
- My parents for instilling in me the values to think differently and to follow my dreams
- The contributors of the case studies who helped to demonstrate the guiding principles with great examples from their experiences (Melissa Brickhouse, Amy Cordova, Rob DeLaubell, John Moleiro, Nick Pettinelli, Chris Richards, Kevin Savage, Brian Timmeny, and Randy Wills)
- Konstantin Nikolaev for taking my concepts and creating clear and valuable graphics from them
- Ginger Levin for reading the book cover to cover and providing insightful feedback and suggestions
- John Wyzalek and CRC Press for providing guidance and support through the publishing process

About the Author

Kerry R. Wills has worked as a consultant and a program manager for Fortune 500 companies on multimillion-dollar technology projects since 1995. During that time, he has gained experience in several capacities: as a program manager, project manager, architect, developer, business analyst, and tester. Having worked in each of these areas gives Kerry a deep understanding of all facets of an information technology program. Kerry has planned and executed over $1 billion of project and program work as well as remediated several troubled projects.

Kerry is a member of Mensa and has a unique perspective on project work, resulting in 10 patents, several published books, and speaking engagements at over 20 project management conferences and corporations around the world. Kerry is a passionate speaker who has a reputation for delivering entertaining presentations combined with vivid examples from his experiences.

Introduction

The business environment has been evolving over the last several decades because of many factors including a volatile economy, growth in technology, increased competition, and demanding customers and shareholders. Companies are also taking larger bets with technology as a means of competitive advantage. The result has been a rise in large and complex programs for many companies. These programs usually span many years, cost millions or even tens or hundreds of millions of dollars to run, involve many people from across the organization and usually also include vendors and partners from other companies.

While programs are getting larger and more complex, at the same time the success rate of programs remains low. In order to manage programs in this environment, it is imperative to take a "consultative approach" to running them. A consultative approach involves managing these programs with a core set of guiding principles and utilizing them across every function of the program. This book defines what it means to take a consultative approach by identifying the eight key guiding principles and then proposing several strategies to enable program managers to be successful in the new landscape by applying them to their programs.

This book's central three chapters explain: the consultative approach in Chapter 2, Consultative Approach and the Program Office; identify the guiding principles in Chapter 3, Guiding Principles; and apply the principles to the program functions in Chapter 4, Program Management Functions. There are also nine case studies that demonstrate how the guiding principles were successfully applied to a program as well as several examples of templates which are described in the book.

Although there is industry material on how to run a program as well as a program office, they are mostly focused on specific functions and tools. This book anchors on the key guiding principles that should be driving the processes and tools, and therefore is applicable to any program for greater success in today's business environment. This book is intended for any professional who is working in information technology and managing a large project or program. It is specifically targeted toward project managers and program managers who have a fundamental background in project and program management principles and want to evolve their skillsets and thinking about how to manage their work effectively.

1

Context and Case

The business environment has been evolving over the last several decades because of many factors including a volatile economy, growth in technology, increased competition, and demanding customers and shareholders. Companies are also taking larger bets and pursuing more opportunities with technology as a means of competitive advantage. The result has been a rise in large and complex programs for many companies. These programs usually span many years, cost millions or even tens or hundreds of millions of dollars to deliver, involve many people from across the organization, and usually also include vendors and partners from other companies.

Figure 1.1 shows a traditional program inclusive of several projects (A, B, and C), with the resources directly managed within the program. These are relatively easy to manage because the program manager has all the control. This model allows a program manager to manage directly without much influence needed.

In contradiction to the traditional program, Figure 1.2 shows the current evolution of programs and how they look today. You can see several changes listed below:

1. The programs are larger and have more projects and work associated with them. As technologies and business solutions become more complex so do the sizes of the programs to implement them. This can be seen with the traditional program example having only three projects but the current program example having several projects within it.
2. The projects are all related to each other and have dependencies. Project A now has a dependency on Project Z, which is outside the program and in another division. Given the complexity and interrelationship of technology, it is common to have a program that has projects which are dependent on other projects in the organization.

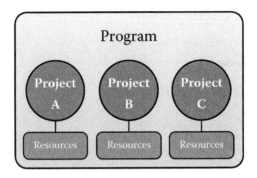

FIGURE 1.1
Traditional program.

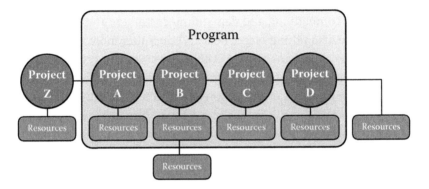

FIGURE 1.2
Current program.

3. Resources are now being used from all over the company. Project B requires resources from a vendor and Project D is using resources from other divisions. Only a few projects today are encapsulated to just the employees from the division running the projects. They usually require other resources, contractors, and vendors from outside the company.

4. Because there are projects and resources beyond the walls of the program, there are also more stakeholders involved and their expecta tions must be managed.

Figure 1.2 represents a visual view of how programs have been evolving and trending. Table 1.1 further expands this view and lists the trending characteristics of programs with specific examples.

TABLE 1.1

Trends of Program Characteristics

Characteristic Trend	Examples
Increased complexity of programs	• Organizations and companies have been investing in technologies for decades resulting in having many systems and therefore many touchpoints to change on new programs. • Strategists are taking bigger bets and pursuing more opportunities with technology as a competitive advantage. • The technology stack is comprised of many different layers, products, and vendors. • The business solutions being implemented are growing in complexity as products and services become more individualized and customer-focused. • Programs can have large structures comprised of many different projects underneath them, many of which have dependencies and interrelationships between them.
Increased duration and cost of programs	• Because of the increased complexity, programs require more resources and time to implement. • It is common today for programs to span multiple years and cost in the millions of dollars. • Costs can include resources, external partners, software purchases, hardware purchases, and vendor contracts.
Additional stakeholders	• As the complexity of programs grows, so does the need to interact with many more internal organizations and external parties. • Communication needs extend to the additional governing bodies within an organization (e.g., audit, portfolio management divisions, finance, etc.). • Additional communications are needed to manage expectations and share information with the growing number of stakeholders.
Evolving resource and skill needs	• To reduce the cost of delivery some companies and programs use offshore resources or shared service organizations. • As the technology becomes more complex, specialized resources are needed, or program teams may choose to build relationships with strategic vendors to help in program execution. • Resources are becoming more "virtual" and working from many different locations including their homes.
Additional controls and guidelines	• New procedures and standards are used to try to manage the investments and improve consistency of execution (e.g., capability maturity model, information technology (IT) infrastructure library). • The focus should be on standards and solutions for the enterprise and not just the program. • Regulations such as the Health Insurance Portability and Accountability Act of 1994 (HIPAA) and the Sarbanes–Oxley Act of 2002 (SOX) with required documents need to be followed.

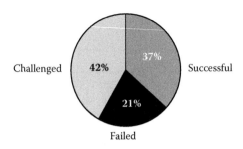

FIGURE 1.3
Success of projects.

Although these trends are realities of today's business environment, what is also true is that the success rate of delivering projects continues to remain low. In 2011 the Standish Group put out their CHAOS Manifesto which surveyed over 10,000 projects. The results found that for projects conducted between 2002 and 2010 only 37% were classified as successful (Standish 2011), which they defined as delivering all the requested functionality, on the expected date, for the planned cost. Figure 1.3 shows the summarized results of the study, which highlights the remainder of the projects that either failed altogether or were significantly challenged.

In order to manage projects and programs in this environment, it is imperative to take a "consultative approach" to running them. Based on my experience, it is not enough just to have a program schedule and manage the people working on it. A consultative approach involves managing these programs with a core set of guiding principles and utilizing them across every function of the program. This book defines what it means to take a consultative approach by identifying these key guiding principles and then proposes several strategies to enable program managers to be successful in the new landscape by applying them.

Chapter 2 explains what a consultative approach is and why it is needed to run programs today. It then defines the charter for a program office and identifies the different models that can be used.

Chapter 3 identifies the critical guiding principles of the consultative approach, for example, having a "single source of truth" for key program information or providing transparency of key information. Each principle is described in detail, and is supplemented with case studies from real programs that span different industries.

Chapter 4 outlines the key functions of a program office but describes them in the context of the guiding principles. For example, in the vendor management function, the subsection on the "single source of truth" explains why having a master inventory of all contracts and invoices is important.

Chapter 5 then summarizes the key points, identifies common themes from throughout the book, and provides some examples of templates discussed.

Although there is a lot of industry material on how to run a program as well as program management techniques, they are mostly focused on specific functions and tools. This book anchors on the key guiding principles that should be driving the processes and tools and therefore is applicable to any program that wants to be successful in today's business environment.

Note that I use the term "program" in this book to refer to an initiative made up of several projects or even a large project. The approaches and principles can apply to either, so I use a loose definition of the word. These concepts can also apply to portfolios of programs as well. Similarly, although the concepts here are focused mostly on technology programs, they can be universally applied to other types of programs and industries as well. Some of the examples and techniques in the book may seem repetitive at times but that is intentional to demonstrate how the guiding principles are applicable throughout all of the program management functions.

This book is intended for any professional who is working in information technology and managing a large project or program. It is specifically targeted toward project managers and program managers who have a fundamental background in project management principles and want to evolve their skillsets and thinking regarding how to manage their work effectively. This book does not dive deep into specific project management techniques, such as how to create a work breakdown structure or a risk log, but instead identifies the critical guiding principles needed to run a program successfully and then shows how to apply these principles across the key functions of a program. However, there are references to key techniques and there are also some sample templates at the end of the book (see Chapter 5, Section 5.3) for the key deliverables discussed.

2

Consultative Approach
and the Program Office

2.1 OVERVIEW

Given the challenges and complexities of running programs in today's environment, program managers have to utilize many tools and techniques to make sure they are managing all of the moving parts of a program including the schedule, financials, vendors, resources, and communications, just to name a few. On smaller projects one project manager can usually handle all of these pieces and does not require as much rigor. However, on programs that have many interdependencies and moving parts, it is critical to understand them all and not just manage them but also have leading indicators that identify trends which can then be used proactively to take action.

Because programs today have a high degree of complexity and have so many moving parts, one program manager usually cannot manage all of the program functions on his or her own. Usually a program management office (PMO or "program office") is set up to help manage the different aspects of a program and provide insight to allow the program manager to manage the program proactively. This book uses the term "program office" to refer to this function.

Because the program office tracks and manages all of the different program functions, this is where taking a consultative approach becomes critical. It is not sufficient to have spreadsheets of data or a program schedule if a program manager cannot use them effectively to understand how the program is trending and where the areas of risk are. Program managers need to have diligence around all of the program management functions. They also need accurate information in a timely manner, which then gives enough lead time to take action and mitigate risks and issues. A risk is something that could have an impact on the program whereas an issue is something that is having an impact.

2.2 CONSULTATIVE APPROACH

I define a *consultative approach* as using consultative techniques and principles to gather insights, identify trends, and positively influence the outcomes of programs. This means using a core set of guiding principles and applying them across the many functions of a program (e.g., financial management, schedule management, and resource management) to understand fully the progress and health of each function. This understanding will provide the program manager with as much information and lead time as available to determine effects and take action if necessary. It is about maximizing the probability of success by utilizing key indicators to make informed decisions. Key indicators can include metrics on schedule tracking or financial tracking such as earned value management techniques.

A consultative approach is needed because of all the trends and program characteristics that were identified in Chapter 1. These evolving program characteristics require much more diligence and transparency than ever before. Table 2.1 outlines some of the evolving needs for the major program characteristics, which can be solved by using a consultative approach.

As a result of the evolution of requirements needed to run programs as well as the current business landscape, the approach taken to run them also needs to evolve. In their classic study on the different bases of power, social psychologists French and Raven (1959) identified five categories of power that are identified and described in Table 2.2.

In the past, program managers had direct authority over project resources and mostly focused on using legitimate, reward and coercive power (see Figure 2.1). Today because most resources are not direct reports, and there are many more external stakeholders to interface with (i.e., Figure 1.2), these techniques prove to be much less effective. Also, because of the growing complexities of technologies and business solutions that are used on programs, it is hard for a program manager to have expert power in either the technologies or the business domains.

TABLE 2.1

Why a Consultative Approach Is Needed

Characteristic Trend	Evolving Needs
Increased complexity of programs	• Having the ability to manage the many different projects, work streams, and components that make up the programs • Understanding dependencies between projects including those external to the program and being able to manage them • Understanding the implications of changes and delays in particular activities to the projects and the program
Increased duration and cost of programs	• Having transparency regarding tracking against schedule and cost objectives • Having the ability to break down large activities into smaller and more manageable pieces • Understanding all of the components that make up the financial forecast as well as trending and analysis of variances • Having the ability to manage milestones proactively and gauge the ability to meet commitments • Understanding the implications of any changes, risks, and issues on the overall schedule and cost forecast
Additional stakeholders	• Having the ability to communicate effectively with stakeholders from different areas who have different agendas and needs • Having the transparency of program health to provide to governing bodies • Using information to manage expectations of stakeholders and explain the implications of specific actions or decisions
Evolving resource and skill needs	• Understanding resource trends and gaps and the impact on the program • Providing insight into key vendor information such as contracts and invoices • Having metrics on strategic partners and their performance against metrics
Additional controls and guidelines	• Having the flexibility to manage the program within the controls and guidelines • Having the ability to demonstrate controls and provide information upon request

In the current environment the program manager must rely mostly on referent power and therefore must take more of a "softer" method toward their interaction with program stakeholders. So instead of telling resources, "Just do it because I am the program manager," the program manager must rely on influencing others and using facts and information to persuade stakeholders and proactively manage trends. This focus on influence, persuasion, and the use of facts is at the core of the consultative

TABLE 2.2

Categories of Power

Category	Description
Legitimate power	Power of an individual because of the relative position and duties of the holder of the position within an organization. Legitimate power is formal authority delegated to the holder of the position.
Referent power	The ability to influence others based on interpersonal relationships and the ability to build loyalty. It is based on the charisma and interpersonal skills of the power holder. A person may be admired because of specific personal traits, and this admiration creates the opportunity for interpersonal influence. Here the person under power desires to identify with the personal qualities of the leader and gains satisfaction from being an accepted follower.
Expert power	The ability to influence others based on one's skills, knowledge, experience, or expertise. It is a function of the amount of knowledge a person has relative to the rest of the team members in the group, project, or program.
Reward power	The ability to influence others based on control over desired resources such as money, gifts, or promotions.
Coercive power	The ability to influence others through the application of negative influence or the removal of positive events. It might refer to the ability to demote or to withhold other rewards. It is the desire for valued rewards or the fear of having them withheld that ensures the obedience of those under power.

approach. Table 2.3 highlights the differences between a consultative approach and a traditional approach to how a program manager needs to run a program.

Given the current environment, including the need for referential power, the traditional approach to managing programs can no longer be effective and therefore programs need to evolve to the consultative model.

2.3 PROGRAM STRUCTURE

One of the most important aspects of a program is how the structure is established. An effective program structure can provide accountability, balances, and checks and enable successful program delivery. An ineffective model can result in role confusion, inefficient use of resources, rework, increases in costs, and missed commitments.

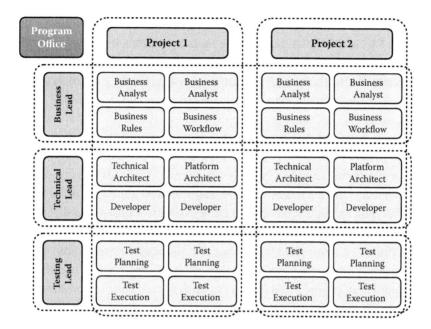

FIGURE 2.1
Matrix program structure.

TABLE 2.3

Comparing a Traditional Program to a Consultative Program

Category	Traditional Program	Consultative Program
Resource alignment	Direct management of projects and resources	Many indirect resources, vendors, and stakeholders
Management focus	Management of a plan	Management of expectations and understanding of which "levers" can be pulled
Plan management	Direct management of resources and plans	Integration of resources and plans that span many areas
Schedule	Find out if the milestone misses the week before the activity is due	Identify trending and being able to "course correct"
Decisions	Based on perception	Informed and based on facts and stakeholder involvement
Information	Lots of data	Insight based on information
Reporting	Current status	Overall health and key trending of risk areas
Vendors	Treat as vendors or subordinates	Treat as partners

The most effective model for managing complex programs is a matrix model. Figure 2.1 illustrates a classic matrix program structure. This program is made up of several projects (project 1 and project 2), which are the "verticals" in this model and are responsible for the execution of their scope. The "horizontals" represent the delivery domains including the business, technical, and testing activities. The program office sits at the intersection of both and drives project standards and project execution.

This matrix creates a natural friction between the projects that need to execute (vertical) and the domains, which own standards and consistency (horizontal). The resources directly align on one area and indirectly on another. For example, in Figure 2.1 the Business Analyst could report to the projects directly but be indirectly accountable to the Business Lead for standard processes and tools. Note that the reporting relationship is not significantly important given the indirect and direct reporting relationships and may also be influenced by organizational culture and standards.

There are many benefits to a matrix model:

- *Accountability.* It drives clear accountability for both project execution and consistency. By having resources directly reporting on one dimension and indirectly on the other, they are forced to consider both priorities as they execute their work.
- *Consistency.* The horizontal model also ensures consistency of execution across the various domains. Without a focus on this as someone's job it may not be considered important and may be disregarded during the project delivery.
- *Integration.* Given how complex programs are becoming it is imperative to have the horizontal functions, which can work across the projects to align requirements, designs, and solutions. Without it, there may be gaps in having everything come together for the program.
- *Scalability.* Because the horizontals work across projects and focus on their domain, this allows for more projects to be easily integrated within the program.
- *Effective use of resources.* Because the horizontals focus on domains, resources can be shared across projects and there is also an understanding of which resources are aligned to what projects and which resources have capacity.

2.4 PROGRAM OFFICE MODEL

Building on the matrix program model, the program office lies at the intersection of both execution (vertical focus) and consistency (horizontal focus). There are several different models that a program office can operate in, depending on how the program is established. The three most common models are outlined below:

1. *Reporting.* This is the most passive model, where the program provides mostly reporting-type functions. This could include administrative support, recordkeeping, and standard reports.
2. *Influencing.* In this model, the program office governs project execution and drives the standards, but does not directly manage the projects.
3. *Managing.* This is the most active model, where the projects and project managers report to the program office, which is directly accountable for the processes, governance, and execution of work.

Table 2.4 summarizes the key characteristics of each program office model, which can be used to determine the right approach to use. For example, if a program office is made up of mostly operational resources and is not heavily process-focused then a reporting model is the best option to use. On the other hand, if a program requires significant rigor and focus on execution, the managing model may be the best choice.

Program office models need to be situational to the nature of the program, skillsets of resources, and company culture. Regardless of the model used, the consultative approaches described in this book are significant ways to demonstrate value and improve the effectiveness of program delivery.

TABLE 2.4

Different Models for a Program Office

Category	Reporting	Influencing	Managing
Project execution	No	Some governance	Yes
Process rigor	No	Yes	Yes
Resource skills	Administrative	Process-oriented	Delivery-oriented

2.5 PROGRAM CHARTER

Once the program structure and program office model have been determined, it is important to consider the charter of the program. This is when the program determines the key objectives of how it wants to operate. There are three primary areas of focus for a program: execution diligence, insight, and a community of practice. Figure 2.2 represents these objectives by showing that the central focus is on diligence of program execution. Next the focus expands to providing insight to enable better management of program execution. Lastly the broader objective is to create a community of practice for the project management professionals on the program team.

2.5.1 Execution Diligence

First and foremost, a program needs to execute on its commitments. This requires diligence across many facets of the program. This means rigorous attention to every aspect of the program with a focus on the objectives and commitments of the program. Areas of diligent focus include the following:

- Planning for the projects and program, which includes creating a detailed program plan and the identification and management of key dependencies within that plan. Rushing through planning activities can result in missed steps, unanticipated costs, and low quality.

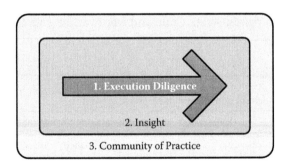

FIGURE 2.2
Program objectives.

A plan is needed to make sure that all activities are identified and planned for, so this needs to be viewed as an investment in planning.

- Tracking granular progress against interim milestones as a gauge of trending and the ability to meet larger milestones. This can also include setting early milestones for the team to build momentum and motivation.
- Managing work intake and change controls to control the entire pipeline of work throughout the program.
- Managing risks, issues, actions, and decisions that influence the different dimensions and outcome of the program.
- Understanding resource needs, skillsets, and capacity across the program. This includes the use of vendors or contractors to supplement key roles.
- Discipline around financial management and understanding the forecasts, financial trending, and variances to plan.
- Focusing on the quality of the work being performed during the program to ensure that what is being delivered meets customers' expectations.
- Providing consistency in reporting and communicating to stakeholders including status reporting and stakeholder meetings.

Having execution diligence regarding the items listed above is the foundation of running any program. Given the complexities of programs today it is imperative to have discipline and rigor in managing these different aspects of a program because one oversight could have implications on program cost, schedule, or resources.

2.5.2 Insight

Building upon diligence, a program needs to ensure that it is providing insight. The Merriam-Webster dictionary defines insight as "the power or act of seeing into a situation." Insight is very different from data collection or information management, and this is an important distinction for a program. Many programs just report on the data, which is not very useful for understanding what they mean and the impacts on managing the program. Table 2.5 shows some examples of the difference between data and insight in the context of running a program.

TABLE 2.5

Difference between Data and Insight

Category	Data	Insight
Schedule	List of activities with dates	Understanding trends toward meeting milestones and commitments as well as managing key dependencies between project milestones
Financials	Report of money spent	Understanding where the money is being spent, having confidence in future forecasts, and knowing what options there are for financial flexibility
Vendors	Tracking contracts and invoices	Understanding the performance of key vendors and the value they are providing
Resources: Human Resources	Program team roster of human resources	Understanding capacity management and identifying the ability to share resources across the program and aligning them with areas where they can provide the greatest impact
Resources: Materials	Program resource inventories	Understanding the full list of program resources and timing of acquisition (e.g., a bill of materials)
Changes	Having a list of changes	Understanding the implications of changes on schedule, cost, benefits, and resources in order to make informed decisions

Table 2.5 shows that having insight is so much more powerful than just collecting and reporting on data. A few ways that programs can obtain insight:

- Having transparency across all stages of the program including work coming into the program (intake), the current projects within the program, and any changes that are being requested.
- Utilizing key performance indicators, metrics, and reporting as a fact base for making informed decisions regarding the management of the program.
- Providing stewardship of information through having master inventories of key aspects of the program such as the project schedule, contracts, resources, and financials.
- Having strong analytics and tracking of key metrics that can inform stakeholders about the progress of the program.
- Utilizing project management and program management tools that usually come with standard reporting and metrics.

2.5.3 Community of Practice

Once the execution diligence and insight functions are established it is also important to focus on creating a project management community of practice. This is especially helpful for a program manager who has a broad organization or program with many project managers who need to work together and use standard processes and tools. The program manager should start by defining a charter for the community of practice that outlines the objectives, operating model, and roles. Then the program manager should outline key activities such as town halls, competency improvement plans, recruiting plans, and surveys.

Creating a community means investing time in building the project management competencies of the team members on the program. Regardless of the tools or processes that are in place it is the project managers running the projects who are involved in the details every day, understand the nuances of the project, and are the greatest influencers of project outcomes and therefore require a high degree of support.

Beyond focusing on competencies, a community of practice can also share practices, templates, and examples and foster a sense of belonging. Project managers can also discuss scenarios and challenges they are facing to get peer feedback and recommendations. Lastly a community can have virtual components such as common documents, community newsletters, or shared books and articles.

There are several benefits to having a community of practice especially on larger programs that have many project managers. Several benefits are listed below that all result in better outcomes for the program.

- Increasing project management competencies allows for an increased performance by project managers and thus a better result on their projects.
- Providing stewardship to project management practices and evolving the processes and tools used on the program aids the project.
- Providing a sense of community allows project managers to utilize their network of peers to get counsel when needed and to share best practices on tools and processes.
- Promoting increased morale and a "feeling of belonging" should translate into enhanced job satisfaction and performance.

With all of these objectives in mind, it is important for a program to define its charter and be deliberate on what its goals are and how it plans to approach them. These should include driving the program toward meeting its commitments, leveraging insights and facts better to manage proactively toward its goals, and building a strong project management practice that can enable success. An example of a program charter is included in Section 5.3.4.

2.6 PROGRAM FUNCTIONS

There are many functions that a program needs to perform well in order to meet its commitments. Figure 2.3 shows these different functions broken into three categories; work intake, project execution, and program operations.

2.6.1 Work Intake

Controlling the "front door" into a program is important because work can come in from many different ways (as shown by A, B, and C in Figure 2.3) including business needs, mandatory or compliance initiatives, other

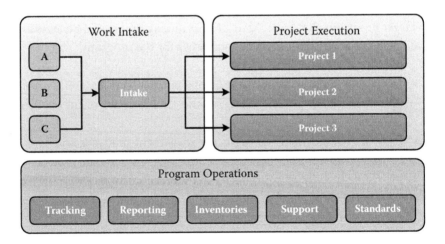

FIGURE 2.3
Program functions.

projects having dependencies, or other requests from within the organization. Having a defined work intake function allows for the program to understand new work and includes the following processes:

- *Assessment.* This entails understanding the request for work including the expectations for scope, schedule, and resources.
- *Governance.* Most programs have multiyear roadmaps that represent key business and technical milestones and dependencies. Governance means determining if the intake request adheres to any technical or business roadmaps or standards. This is an especially important control mechanism if the program spans many years and has an elongated roadmap. Short-term tactical decisions could have implications for longer-term strategic roadmaps if not governed properly. For example, a short-term code update may add more complexity to a technology component that was planned on being retired, thus making it harder to retire.
- *Estimates.* This means providing initial estimates of cost and resources.
- *Capacity.* This means determining if the program has the resources to conduct the work and committing them to the initiative.

2.6.2 Project Execution

Once work has come through intake as is understood, the projects are then initiated and executed. This is where project managers deliver their projects against business commitments throughout the software delivery lifecycle from initiation through to closing. During this time, project managers can utilize many foundational management techniques including schedule management, resource management, financial management, and vendor management.

2.6.3 Program Operations

Supporting both work intake and project execution is program operations. This role is usually performed by the program office and Table 2.6 outlines some of the key functions of program operations.

In order to successfully manage a program, the program office needs to consider all of these functions from intake to execution as well as all of the operational activities that support it.

TABLE 2.6

Program Operation Functions

Function	Description and Value	Examples
Tracking	Monitoring key program metrics to understand progress and ability to meet commitments. This information will be used to manage the program and make informed decisions to mitigate risks and issues.	• Program schedule and milestones • Key dependencies within the program and with other programs • Financial forecasts and spending across all cost categories (e.g., resources, vendors, technology) • Resource management including resource alignment to projects, resource skillsets, open roles, and resource forecasts • Tracking changes to projects with their effects on cost, schedule, and resources • Highlighting program issues and risks with their impacts and path to closure
Reporting	Aggregating information from projects across the program that will allow stakeholders to understand the health of the program.	• Program health as measured by the status of each of the projects across several dimensions including scope, schedule, resources, and budget • Financial reporting including variance analysis, forward-looking forecasts, and spending to date • Dashboards and status
Inventories	Having a master inventory as a single place for housing key program information. Master inventories avoid confusion and provide clarity of critical information.	• Project list and description • Program and project scope • Vendor inventory with contract and invoice information • Communication plans including key stakeholder meetings and communication methods
Supporting	Supporting the program administration.	• Bringing new resources into the program and getting them up to speed • File management • Calendar and meeting management
Standards	Providing standard tools and processes for managing projects within the program. This provides consistency and the use of best practices.	• Project management tools to manage schedule • Estimation tools and processes • Status reporting • Templates and other project deliverables

3

Guiding Principles

3.1 OVERVIEW

Chapter 2 explained the reasons why a consultative approach is needed to run programs today. Because of the nature, size, and complexity of programs today it is imperative that a program focus on all three objectives outlined in Section 2.5, which are execution diligence, insight, and a project management community of practice. Within each of these objectives is a core set of guiding principles that make up the consultative approach. Figure 3.1 shows these eight guiding principles in the context of the objectives.

- Execution diligence means staying on top of all the activities in a program (diligence), paying attention to the specifics as well as managing and presenting information in a simplified manner.
- Insight includes providing transparency of progress and information, using that transparency to facilitate fact-based decisions, and keeping critical program information in one authoritative place.
- Creating a community of practice with skilled program team members requires focusing on leadership, stewardship, and ownership (the "ships") and a focus on the internal and external customers.

The focus of this chapter is on explaining each of the guiding principles, identifying techniques to using them, and then sharing a case study that demonstrates how the guiding principle was used on a real program.

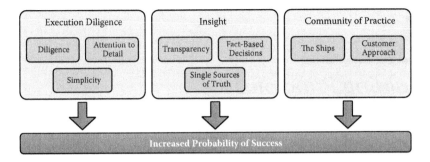

FIGURE 3.1
Guiding principles for a consultative approach.

3.2 DILIGENCE

Diligence is about staying on top of the many moving parts of a program and is also described as simply working hard. This is especially important given the complexities of programs today and with the many account-abilities that are intrinsic to the matrix model described earlier. Because diligence is a foundational aspect of managing a program there are many benefits a few of which are noted below.

- It allows program managers to keep watch over the many program activities, enabling them to understand risks earlier, identify any trends, and articulate progress to stakeholders more effectively.
- There are so many different aspects to managing a program that losing focus on any one item can mean an unexpected issue or an impact on the schedule or cost. Diligence enables the ability to find issues earlier and thus minimize their effects because it provides time to remedy them.
- The entire team relies on the program manager to bring structure and order to the program. There is a compounding effect when a program manager is not organized in that the team becomes less effective as well because they are not clear on the work.

There are many examples of where diligence can be used on a program:

- Making sure that all project schedule activities are tracking to mile-stones; current deliverables are on track for their dates and any new activities start as expected.

- Ensuring that resources are available when needed and with the appropriate skillsets required.
- Tracking that the program financial forecasts and spending align with expectations of the budget. For example, comparing actual cost spent to date against an expected spend of budget would show if the program is tracking properly.
- Ensuring that vendor contracts are up to date and relate to the work required of the contractors.
- Understanding changes and then determining what to do with the changes and making the appropriate updates.
- Monitoring any action items, issues, or risks; having dates assigned to them with the owners and confirming that they are tracking to those dates.
- When a problem does arise, identifying the actions and driving them to closure as soon as possible.
- Staying on top of all the operational and logistical activities on a program including meeting scheduling, bringing on new resources, and document management.

Diligence can also include proactively managing activities such as looking ahead at contract expirations, resource start dates, future forecasts, and upcoming purchases to make sure they are all tracking to the expected plan. Usually it is the program manager who has to keep all of these items in mind and track them, so if the program manager is not staying on top of all of these items then most likely they will not get the attention needed and possibly will not be completed on time or accurately.

3.2.1 Techniques

Diligence is about staying on top of many items on a program and there are several methods for doing this important function.

- *Define and document the program management approach.* Early in a program, it is important to define and document the approach for managing the different operational components of the program. Most standard delivery methodologies have a program management approach or charter deliverable that can be used. The format or template does not matter as much as the fact that the program manager thinks through, plans, and communicates how the program will be managed. For example, this document can include the approach to track financials, vendors, and resources as well as key communication and reporting mechanisms.

- *Understanding the critical path.* The critical path is the sequence of project activities that aggregate to the longest duration of the project based on dependencies. The critical path represents the quickest possible time to deliver a project. A project can have several concurrent streams of work but any delay of an activity on the critical path will have an impact on the project schedule. Understanding the critical path of the projects and between the projects in the master program schedule will help a program manager understand where the risks and issues have the biggest impact on the program.

- *Have rigor around the plan and upcoming activities.* The current program landscape has many moving parts that the program manager needs to stay on top of, especially on programs with multiyear roadmaps. Because program roadmaps can span many years, having rigor around individual activities and milestones will ensure progress toward the long-term goals one milestone at a time. Not managing these activities will lead to what I call the "pile phenomenon" where there are piles of work to be done that continue to grow. It is difficult to manage a program once the piles start to grow because everything becomes reactive to the piles as opposed to getting in front of the work. Remember that there is no activity on the project schedule called "later" and therefore pushing activities off until "later" is not optimal for success. Not only should the program manager focus on in-flight activities, but he or she should also spend time looking at the upcoming activities to make sure that they will be ready when the time comes or to see if some work can even be started early.

- *Plan for small units of work.* Because of the trends identified earlier in this book, projects and programs are growing in complexity and size. As a result milestones on projects are getting farther apart because of longer durations to deliver solutions. A project manager needs to have a clear idea of the status of the project work to gauge the ability to meet commitments. Therefore it is a good practice to plan for small units of work that can be managed tightly. Then the program just becomes an aggregation of these small units.

- *"If it is not a list then it does not exist."* Every outstanding activity should be captured on a project schedule or in an action item log. These documents need to be "living" documents in that they are constantly being reviewed, updated, and used to track and manage the progress of the activities.

- *Be a professional harasser.* Once the items are on a list with a date, the program manager or project manager then needs to follow up as the due dates are approaching to confirm they will be done on time.
- *Leverage tools.* Many project management tools and even file management tools have workflow components that can send out reminders, follow up on status, and capture key information. These should be considered as another way to track and manage program activities.
- *Have effective meetings.* Meeting management is a critical aspect to staying on top of activities. Meetings are important because they are where milestones, issues, and risks get discussed as well as where key decisions get made. Preparing for meetings in advance by determining what actions or decisions are critical to have will allow the meeting organizer to keep the group focused and drive for closure. Having effective meetings also means facilitating conversations properly by keeping the team members on track. It also means capturing key takeaways and ensuring that they get completed on time.
- *Use a checklist.* For standard processes, checklists are a great way to make sure that all key activities are recognized and tracked. For example, when a new project starts, a checklist would be a good way to ensure that the proper operational activities happen (e.g., creating a status template, adding the project to the master project inventory and schedule, assigning resources, etc.).
- *Establish clear accountability for deliverables and results.* To ensure diligence on program activities, a program manager needs to hold team members and leads accountable for specific deliverables and results. This means having clear definitions of responsibilities so team members and leads are aware of those deliverables for which they are accountable. Overlapping or unclear accountabilities allows for finger pointing, missed expectations, and confusion on the program. This can be accomplished through documenting roles and responsibilities and ensuring named resources on program and project schedule activities and action items. Tools such as a RACI, which explains the four types of accountabilities including Responsible, Accountable, Consulted, and Informed; or Responsibility Assignment Matrix (RAM) can be used to document roles and accountabilities.
- *Be decisive.* One of the biggest challenges on programs is that project managers and program managers do not feel that they can make

decisions. Part of being diligent and keeping the project moving is to make decisions in a timely manner. Having too many decisions pile up will cause delays and a loss of confidence from the project team. The key is gathering as much information as possible including input from team members and then making fact-based decisions or a "best professional opinion" based on known information. A program manager needs to recognize the importance of making quick decisions supported by facts and the team's input. Not every decision will be right, but sometimes action is better than inaction. It will be a judgment call for the project manager as to when she can make a decision and how best to make it.

- *Provide structure where there is none.* Many times program information or activities are not well organized and therefore resources are not being optimally effective. It is up to the program manager to provide this structure in the absence of it. This can include crafting information into a story for a presentation, providing a template to collect information that will be analyzed to solve a problem, setting up meetings to resolve issues, or just organizing the work into a project schedule.

- *Trust but verify.* Ronald Reagan was famous for repeating this old Russian saying, and it means that although a source of information might be considered reliable, one should perform additional research to verify that such information is accurate or trustworthy. This is true for all program information, and it is the responsibility of the program manager to ensure that the information is accurate because decisions and actions will be based on it. Staying on top of activities, having the right performance metric such as schedule or financial tracking, conducting peer reviews, and having open discussions with team members are all ways to verify information.

As the techniques above demonstrate, diligence is about proper planning and managing of all work from the program level to the project level and even to the smallest level of operational activities.

Case Study: Using Diligence to Manage Program Cost Variance; Telecommunications Industry

(Contributed by Brian Timmeny)

A management consulting company was brought in by the executive team of a global telecommunications leader who had recently completed a merger/acquisition. The telecommunications company had purchased a unit within the European marketplace and was now looking to integrate that asset fully. Upon the close of the purchase, the company now found itself owning two disparate systems to manage customer service relations and integrated financial reporting.

The company faced the complexity of two units that were unable to report holistically against key segment initiatives. This left each unit running delivery and operational overages. In many cases, these overages were magnified due to the reality that management attempted, although unsuccessfully, to manage across two distinct systems. This not only made customer service and financial impacts difficult to track, but made reporting against items across organizations a near impossibility. A program was initiated to combine the two organization systems.

Strategic program planning began late in the year, with a program goal of delivery kickoff early the following year. The company's team brought the consulting organization into reviews where each unit highlighted the losses that were occurring within the current program delivery teams, as well as the opportunity cost losses incurred by not holistically combining the two organizations' customer service and financial reporting domains. The executive team recognized the challenge of inconsistent delivery within the two operations, and understood the need for program diligence and program visibility metrics.

EARLY PROGRAM DAYS AND CHALLENGES

Early in the consulting team's joining the program, a few key items were noted around the delivery team's diligence. As the wider delivery picture began to take shape, some of these early challenges included:

- Intake and change control existed as separate processes.
- Estimates were rarely quantified in a measureable or consistent method.
- There was limited understanding of the detailed modules the program was responsible for implementing.
- The program team faced a monthly overrun of nearly $2 million ($20 million since program inception).

Although the wider executive and program management team agreed on the need for elevated diligence and heighted scrutiny against the program's delivery, financial reporting remained an open issue.

Many of the executive team and program management team members had grown accustomed to the current unit-based financials. This method of financial reporting gave little insight into the true nature of delivery against program commitments. In order to obtain a more accurate view against delivery, two key steps were taken:

- Consistent implementation of milestones, as measured against consistent module definitions
- Expenses by individual cost center to enable asset/module level reporting

BUILDING THE DILIGENCE FRAMEWORK

With agreement between the two organizations regarding the need to define visibility in a cohesive way, both the program and delivery teams were able to decide upon concrete next steps in order to set in place the foundation that would allow for long-term visibility, and begin to force a culture of delivery diligence. Below are some of the items implemented toward that end:

- Identified a comprehensive list of assets/modules that would be implemented
- Mapped assets against the capabilities they would enable
- Mapped the capabilities against the business objectives they would enable
- Created the schedule domain, with milestones attached to each phase of asset delivery
- Implemented a financial model, utilizing consistent cost centers, to report at the asset delivery level

With the business objectives, capabilities, assets, schedule information, and financial information now implemented within a common repository, the team now had the ability to cross-reference data in order to view detailed program information from several pivot points.

IMPLEMENTING DILIGENCE

With a foundation of common schedule tracking, and financial discipline by asset, the team now had open to them an ability to move forward on the several required data pivots that would show a holistic view of program delivery. Several key documents formed the early basis for holistic program review:

- *Asset roadmap.* Reviewed future module states by individual application
- *Capability roadmap.* Showed the realization of capability within a time horizon
- *Business objective roadmap.* Viewed the business objective delivery timeframes
- *Program roadmap.* Allowed for a comprehensive review of the program delivery

With the ability now to understand the schedule and cost impacts across each of the critical pivots of business objectives, capabilities, assets, and programs, the team now had tremendous insight into the interdependencies of the views within their projects. Each of these views had the ability to pivot, and give insight into, one common delivery and implementation. The team could now understand where they stood in relation to each critical pivot point, and informed decisions (schedule, cost) could now be made against program priorities in response to market and budget conditions.

IMPLEMENTING AND MANAGING CONSISTENCY

With the framework now firmly in place to understand the impacts of decisions upon program, asset, capability, and ultimately business objective delivery, the team now set to the task of implementing consistency against program reporting.

The extended team worked to define the key levels against which reporting would be implemented, known as reporting views. Each of these views pivoted around the uniformity of data surrounding schedule, cost, scope, and risk. By reporting on each of these items in detail through the executive level, items would be understood in relation to each other across delivery teams. The initial team to create these standards sat between both organizations, and acted as an independent team, so as to create consistency between both organizations.

Program status reports were the first documents implemented. These multiple-page documents gave a pivot out of the program schedule data, cost views (by asset), scope, or risk. This gave a comprehensive view of each project, inclusive of risks, accomplishments, and upcoming milestones.

Program status served as a roll-up of the many projects that made up the engagement. This view allowed for a higher-level review against project schedule variances, cost shifts, and cumulative project risks. Within three months of implementing these reports, although schedules continued to vary, cost estimates began to normalize to within 5% of average estimated budgets (this remains significant in that the estimation process had not yet been fully revised by this point in the program).

The final suite of reports implemented was administrative in nature. Due to the fact that data could now be reported from one common repository, the understanding of program thresholds could now be calculated in a short timeframe. As variances would arise, the program team could now review these data against agreed-upon metrics, and alert the associated program manager or delivery manager. These administrative reports would allow for a detailed review of ongoing metrics and ensure near-realtime information that managers (program or delivery) could then act upon in a timely manner.

With the implementation of these key elements toward diligence, several governance activities were performed:

- Definitions of gate and milestone quality criteria were determined.
- A central delivery review organization tasked with code and test audits/reviews was formed.
- Centralized work intake and change control teams were established, and weekly reviews scheduled.
- Monthly governance/program reviews were established.

With these new implementations, critical decisions were now made with a comprehensive view of delivery. An early decision (relative to the datastore stand-up) was made to hold delivery of a credit card payment module, as it did not align to the key program objectives. This decision alone saved the program nearly $1.5 million (including credit card vendor fees) during the program's second fiscal year.

THE RESULT OF DILIGENCE

Within three months of the process implementation, the monthly reviews had resulted in an average monthly spend of $2 million, with clear data pivots (financial details) against projects, assets, capabilities, and business objectives. The first quarterly review had not yielded full program guardrails yet, with a program overrun of still nearly

$400,000; the power of diligence, in this review, showed that the over-run was related to only two of the 28 assets within the scope of the pro-gram. The second monthly review left the program, however, with a $100,000 underrun (leaving the program with a two-month $300,000 overrun). The third month yielded a $140,000 underrun, bringing the first asset fully back within schedule and budget standards (leaving the second asset with a net $160,000 overrun).

Program variances continued. However, the diligence now offered by the implemented framework gave the program and delivery man-agement teams effective tools, through holistic visibility, to determine and implement appropriate program solutions. The framework did not provide a one-stop solution for the program, but rather enabled capabilities around visibility and insight that would allow the manage-ment team to better react and devise alternatives to ensure a program managed tightly within the guidelines set down and, more important, enforced by the team.

3.3 ATTENTION TO DETAIL

Accompanying diligence is the principle of "paying attention to the details." If diligence is about providing structure and staying on top of all the activities, then attention to detail is about understanding the specifics and making sure that the quality of the work is accurate and complete. There are many places on a program where paying attention to the details is critical. The examples below are organized into the two categories of understanding the details and then focusing on their quality.

3.3.1 Understand the Details

- Knowing the specifics of the projects in the program. Although it is impossible for a project manager or program manager to know everything, he should know enough of their scope to be able to dis-cuss the project with stakeholders as well as articulate the implica-tions of any changes or risks.
- Knowing the schedule of work with deliverables and key dependen-cies so as risks and issues arise there can be fact-based decisions based on these implications.

- Understanding the details of a contract with a vendor can mitigate risks if a conflict arises, and the legal teams need to reference that contract.
- Recognizing when there are resource constraints or gaps that need to be filled by understanding who is on the team, what skills they possess, and what resource needs the upcoming activities require.
- Conducting a thorough analysis of a change, issue, or risk to understand the specific details and implications of those items so that options can be identified and informed decisions can be made.
- Identifying and managing all of the action items and decisions that need to be tracked on the program.

3.3.2 Ensure Detail Quality

- Making sure that numbers add up and that the math works on financial calculations. One wrong decimal place or zero could have implications of thousands of dollars if incorrectly misrepresenting financials.
- Confirming that all action items, issues, risks, and plan activities have dates and named owners associated with them so they can be assigned and tracked.
- Ensuring the most effective structure and story of a presentation to stakeholders. This will improve their confidence in the program and influence them to make key decisions. Do not underestimate how a messy presentation of work can trump the quality of the content.
- Making sure that there is traceability between program deliverables. There are usually many different documents on a program, and it is important that they all are consistent. The team loses credibility and cannot be on the same page when several documents have opposing information.

During my time as a consultant I heard the phrase "quality at the source" many times. This means you own the work and should not assume that someone else will find the problems. One common example during a program includes a time when developers do not check their work because they may assume that testers will find the defects. The point is that each person has to view his work as a representation of his own reputation and the program, and that the quality that is presented demonstrates how much each person cares about that. It also demonstrates taking ownership of one's own work and not letting "someone else find the mistakes."

3.3.3 Techniques

Paying attention to detail is a mind-set and also requires an investment in time to get into the details.

- *Invest the time.* Block off time during the day to spend on reviewing details or performing quality reviews of documents. This has to be viewed as an investment in time to understand the details of the work fully. This time can include having team members walk through the specifics of the program solution.
- *Check your work.* Yes, we learned this in grade school, and it sounds simple, but I am constantly surprised by the quality of some presentations that are made to large audiences of senior stakeholders, let alone things as small as documentation or e-mails. The onus of quality and accuracy needs to be on the team member producing the work and not on the recipient of it.
- *Find good examples of high quality.* Look at deliverables from other program members or other programs within the company who are seen as producing high-quality work to understand some of the techniques they use and how they approach the structure of their work. Ask them to explain their approach and the key considerations. This is a great example of something that can be facilitated through the project management community of practice described in Secton 2.5.3.
- *Conduct peer reviews.* Having a peer review documents is one of the most effective ways of ensuring quality of deliverables. It is always helpful to get a fresh perspective on one's work, especially from someone who is familiar with the project. Examples of this include developers performing code reviews of other developers or a business analyst reviewing a peer's requirement documentation before going for business signoff. Peer reviews should be informal in nature.

Case Study: How Paying Attention to Details Helped a Payroll Billing Project to Be Successful; Financial Services Industry

(Contributed by Randy Wills)

While working as a director of operations for a large property and casualty insurance company, I was tasked with the responsibility of being the business lead for a new billing option for our customers.

Historically for workers' compensation products, insurance companies bill their insureds based on their workforce's payroll information obtained at the point of sale. However, if an insured's workforce fluctuates during the year this is only trued up at the end of the year via an audit. There was great demand to build a pay-as-you-go option with our insureds to manage their cash flow better based on market conditions and worker fluctuation. Thus an internal project was kicked off to look into building a self-reporting pay-as-you-go billing system based on user entry of their payroll.

Having worked as an IT project manager, a business project manager, an agile coach, and consultant, I have played many different roles on a given project and have found that being a successful lead or project manager is very reliant on that person's ability to pay attention to all of the details. This project was one where the team was successful in implementing a new billing function in under one year including a broad range of IT changes, process changes, legal/compliance approval, marketing materials, and associated metrics. My belief is that the success of this project was due to the core team's ability to pay attention to all of the details starting early in the life cycle. Two examples of how we managed the details are outlined below:

1. Real-time updating of risk and issues logs. Having a common place to post questions, action items, risks, and issues that are updated frequently will lead to more accuracy and rigor around the project.
2. Detailed requirements. Thinking through all of the affected functionality, then building out scenarios, and finally asking detailed clarifying questions to drive out all possible outcomes and requirements is critical to ensure accuracy, quality, and completeness.

REAL-TIME UPDATING OF RISK AND ISSUE LOGS

This project had many different teams working on it including billing, IT systems, commissions, policy output, audit, sales, marketing, call center, and field operations, and thus it was important that as discussions were had, things such as action items, issues, and risks were documented in a central location and addressed appropriately and in a timely manner. The team used the action item, issues, and risk logs to capture items when they occurred during the meetings. By having a central location to store and document status on open items, it allowed all members of the team to stay on top of any remaining tasks and items not resolved.

During the week the project manager would follow up on any open action items, issues, and unmitigated risks so that when the weekly team meeting took place, we could review and discuss where we were on the project. What made this successful is that the project manager did not view this as a weekly activity where he would just ask for status on the open items at a status meeting, but viewed it as a living document that was to be updated in detail daily (and even hourly) as meetings took place and project material was discussed. The documents were updated in detail, such that any outsider could read them and understand the specific intent of the information. This made the documents more than just based on a point in time, but a fluid view into the project's health. The lesson here is that these logs are only as good as their maintenance and that updating them after meetings and as conversations take place will make the logs more pertinent and useful and not something simply to read out loud to everyone during a status meeting.

DETAILED REQUIREMENTS

Often times on projects, the business gives the "intent" of the requirements, and it is incumbent upon the IT team to draw the specific requirements out of the business through questions and options. This project was no different in that the business started off with a good understanding of the "intent" of the project, which was to build a website that an insured could key payroll into directly. What made this project a success was the collaboration to take the requirements to the next level of detail through iterative questioning and providing screen mockups to ensure everyone had something tangible to review and work through.

During the initial few weeks of requirements, the team spent time building out all of the affected functionality that we could think of to document what was labeled the "capability" matrix. Everything was considered from the intake process to the website flow, to the output of the process, to what we would provide the call center when calls came in regarding the new payroll billing functionality.

Once we had a complete set of functional effects in the capability matrix, the team then scheduled many requirement sessions to work through the details even further. With many affected areas, this became difficult but necessary. What helped to facilitate the discussion was when the business analyst pulled together detailed specifications ahead of time from the key business owner of the functionality that was being developed. For example, if the upcoming session was on the user interface of the new payroll entry system, then she would meet with the billing lead to detail out a first pass at the requirements. This way there was a starting point to the requirement sessions, as opposed to walking into these sessions with a blank piece of paper. Having something tangible proved useful as a common starting point where the business analyst could then ask questions such as, "What is the happy path of the user entering payroll?" or, "What if the user wants to add a new state to the policy?" or, "What if the user wants to start and come back later?"

The lessons learned here are to start at the beginning by documenting all affected functionality, then build a starting framework by meeting with the subject matter experts first, and finally bring together a broader group to start working through the details and scenarios. The iterative nature of the requirements is often why projects take very long during this phase of a project, but this is where the commitment and level of detail are most important as they set the tone for the quality, accuracy, and functionality of the project's outcome.

CONCLUSION

By maintaining the central project logs and drilling into the specifics of the requirements, the team was able to identify and manage many of the detailed aspects of the project. This gave us good insight into the project goals and how we were tracking against them, which ultimately resulted in a successful outcome for the project.

3.4 TRANSPARENCY

The guiding principle of transparency is focused on gathering information and organizing it in such a way that allows stakeholders to understand themes, trends, and risks to the program as early as possible. This aligns with the objective of insight discussed in the prior chapter. This is critically important to have so that, as program managers, we have as much time as possible to take remediating action when a risk or issue arises. The earlier that a trend is identified, the more time a program manager has to take corrective actions. The only way to get these early indicators of progress is through transparency of program information. Several examples of this are highlighted below:

- Having clear definitions of scope and what deliverables are expected to be completed on each project
- Tracking the ability to meet program schedule commitments well before the milestones are due through monitoring of the progress toward granular milestones and deliverables
- Understanding financial trends and how the program forecast is aligning with budget objectives
- Understanding resource needs aligned with deliverables and milestones and comparing them to available capacity
- Managing contracts and invoices from vendors to identify where the work is being performed and by which companies
- Identifying and communicating risks and issues to stakeholders with the associated plans for action
- Using program metrics such as defect rate or rework to provide insight into the quality of the program deliverables

Transparency means having a clear understanding of program health and highlighting risks and issues early as well as communicating all of this to stakeholders. In some corporate cultures a yellow health indicator is not viewed in a positive manner, but it is important to note that program teams need to be transparent about where the risk areas are so that action can be taken. Showing a status of yellow with a clear action plan is much better than having the "I can handle it" attitude where status is green for a while and then turns red once there is a realization that it is not being handled well. By the time it is recognized and reported, it is too late to take any remediating action.

Obviously transparency is a by-product of having diligence and attention to detail. The only way to know key performance indicators of a program is to diligently manage the activities as described in Section 3.2 and then to understand the details enough to interpret the information and trends.

3.4.1 Techniques

There are several techniques that can be used to provide transparency across the different dimensions of a program.

- *Track work at a granular level.* By managing work at a low level of detail (e.g., scope elements, schedule activities, financial buckets, interim milestones, etc.) it is easy to see trends and gauge progress toward larger measurements of cost and schedule. It is also easier to aggregate the granular information into different categories based on stakeholder needs of reporting and reviewing the information.
- *Implement a control system.* A control system can measure progress and assess health. "A project monitoring system involves determining what data to collect; how, when and who will collect the data; analysis of the data; and reporting current process" (Grey and Larson 2005). A control system can be used to generate realtime health of a program so that the information is as up to date as possible. A control system does require that its users update information accurately and on a timely basis so it is only as good as the input that is going into it.
- *Utilize status reports.* Status reports are good ways to generate transparency of risks and issues for program stakeholders. This is where health colors can be turned yellow, or risks and issues are highlighted along with their action plans. This can also be used to manage expectations around project delivery.
- *Utilize dashboards.* Programs can generate dashboards that highlight the health of programs across many dimensions. Because programs today are made up of many projects and workstreams, it is helpful to have a single site where stakeholders can go to understand the health of each of these components. A typical program dashboard shows the health of all its projects across the following dimensions:
  - *Scope.* If there are risks to meeting the identified scope of the project, or if the scope is unclear.
  - *Finances.* Compares the current forecast (including actuals incurred) to the baselined budget.
  - *Resources.* Identifies if all resources required to meet project commitments are available when needed. This can include internal and external resources (i.e., vendors).
  - *Infrastructure.* Identifies if any hardware and software purchases are tracking to plan. Because purchasing these items usually

requires a significant lead time this is often an important dimension for reporting.

- *Quality.* Shows key indicators of quality such as testing defects.
- *Benefits.* Tracks the realization of program benefits.

• *Earned value management.* Many programs use a technique called earned value management to manage discrete pieces of work. The concept of earned value management is simple: projects "earn" work as it gets completed, which is then compared to what was planned to be completed (also called "burn") to measure progress. Earned value management requires that cost of work in progress is calculated, which requires work to be broken down into small entities. Although this technique takes some time to plan for, it is effective in helping a program manager to understand its progress. There are significant documentation and several tools in the marketplace that can help with earned value management; an example is demonstrated in Figure 4.6, Section 4.4.3.

• *Clear escalation path.* Related to transparency is having a clear escalation path for raising issues and risks as trends get identified. This allows for rapid decision making and action to mitigate the risks and issues. For example, if one of the program dashboards is showing that a project is at risk of not meeting a milestone there should be a documented process for whom to escalate to, when, and at what meeting/format.

Case Study: Transparency during a New Product Development Program; Government Sector

(Contributed by Rob DeLaubell)

The notion of transparency in managing projects and programs exists in various disciplines, such as schedule management (e.g., Will we deliver when we committed?), financial management (e.g., Will we deliver in line with budget?), and vendor management (e.g., What are the roles and responsibilities between employees and consultants?). For this case study, I refer to a new product development program within Personal Line Property and Casualty Insurance, where we developed a new homeowners' rate plan, and I cite examples of successfully using transparency in program management. The program team consisted of over 100 IT resources at its peak, spanned over three years from inception to rollout completion, and cost tens of millions of dollars. My role was the lead IT program manager.

TRANSPARENCY IN SCHEDULE MANAGEMENT: WILL THE PROGRAM TEAM DELIVER WHEN COMMITTED?

The key idea here is to derive and agree on the incremental milestones and key dependencies with program stakeholders (oversight/steering committees, business, as well as the IT program team) from program start through final delivery. My program had eight different IT work-streams, each having its own development team, project manager or lead (full or partial allocation), and associated schedule or plan.

At the start of the program, each workstream provided its key mile-stones and cross-team dependencies, which were denoted within their respective plans using an agreed-upon format. The milestones and dependencies were compiled—a mostly manual process (a one-time deal)—into a program-level milestone plan using Microsoft Project, with one column containing an originating workstream. Program-level content also lived in this plan as well (cost-benefit analysis [CBA], governance gates, etc.). There were several hundred milestones repre-sented overall, given multiple releases supporting a state rollout.

The weekly reporting and monitoring rhythm consisted of applying predefined filters on the program milestone plan to extract milestones past due, due this week, and due within the next two-week window. This view allowed workstream project managers to speak to status and variances, and called out risks and in some cases issues. It was my experience that patterns emerged over time in terms of which work-streams consistently achieved all milestones versus those where there were schedule challenges, and thereby provided another datapoint in terms of where to focus my time and attention the better to enable the teams to be successful. In some cases, corrective action was necessary to address variances. For example, in some cases subsequent discus-sions resulted in schedule updates to workstream plans where addi-tional time was necessary (milestone dates remained unchanged in the program-level milestone plan, with variances noted). In other cases, higher-level scope discussions were necessary with business, whereby scope was deferred to later releases.

In many cases, a milestone deliverable from one workstream was a dependency for another workstream. A typical example is quality assurance (QA) entry (and within my organization, there are metrics associated with meeting defined QA entry dates). I recall more than one agreement with QA for a phased QA entry in order to mitigate schedule delays. I do believe that the transparency created as a result of the milestone plan resulted in as early a read for such delays as possible, allowing time to craft a mitigation approach.

In practice, the value of having the milestone and dependency view was an effective tool to increase confidence in delivery and maintain transparency to program stakeholders. Perspectives arising from the end-to-end plan resulted in many productive discussions with business stakeholders during the execution of the program. Although not all milestones were met, the program-level plan did provide early reads, and brought focus to areas requiring re-planning.

TRANSPARENCY IN FINANCIAL MANAGEMENT: WILL WE DELIVER WITHIN BUDGET?

Forecasting is a never-ending process. The goal is always about "closest to the pin;" in other words, it is equally as "bad" to have a large favorable variance as it is to have a large unfavorable variance (a large favorable variance over the year is problematic because capital was allocated for the program that could have been invested in other programs, thereby yielding a rate of return). One of the tools that I used to ensure the program was tracking within +/–10% of forecast was to track program spend on a weekly basis using an over/under report. This report compared time charged to the program against forecast allocation. Using time as a proxy for dollars, I was able to identify variances developing during a fiscal month, and apply corrective action. I routinely filtered out the minor variances (the "noise") to show only resources with more than 10 hours variance and more than 10% variance. Below are some specific examples of patterns that I encountered on my program, the corrective actions taken, and benefits of having this level of transparency:

- *Resources overcharging.* Need to increase allocation: resource may have been allocated 50% to the program but in reality the need was greater. Corrective action in some cases was to increase forecast for future months; in other cases, this was an early indication that scope or complexity was greater than expected, which may have required broader schedule/scope discussions with program stakeholders.

- *Resource overcharging.* Incorrect program charge: in some cases, resources should not have been charging the program. Corrective action was having the resource correct the timesheet.
- *Resource undercharging.* Resource unable to engage in program: concern here is that program work is not being done. Perhaps a resource is unable to spend time on program work due to demand from other areas, an early indication of schedule slippage, in which case corrective action is a discussion with appropriate resource managers, or other project managers, to either enable resource's time, or secure another resource with needed capacity. Alternatively, perhaps work cannot begin due to a dependency, which again requires intervention.

Another point of transparency within the financial space is the notion of tracking dollars to "approved spend." However, approved spend may be a changing figure, due to periodic (e.g., monthly) approved change controls. I maintained a log showing incremental approved spend—essentially a financial storyline starting with approved CBA dollars through all currently approved change controls—and it was a standing point of communication/discussion at oversight meetings and steering committee meetings.

TRANSPARENCY IN VENDOR MANAGEMENT: WHAT ARE THE ROLES AND RESPONSIBILITIES BETWEEN EMPLOYEES AND CONSULTANTS?

I have found in this and other programs, that having discussions with vendor peers (project manager [PM] to PM, tech lead to tech lead, etc.) around expectations and accountability is fundamental to successful delivery. It is critical that folks "on the ground" have a common understanding of contract language. I have used RACI diagrams to help document these, and have always found that equally as important as the final RACI artifact itself is the common understanding that is achieved during the course of its development. Accountability then becomes transparent to the program team (both employees and vendor partners), project stakeholders, and project oversight.

CONCLUSION

The above points will not guarantee success, however, they will position program teams to be successful. The absence of practicing transparency in program management will lead to execution issues such as cost overruns, schedule delay, or lack of clarification in terms of accountabilities. The program cited in this case used transparency of schedule, finance, and vendor information to gather insights and then provide lead time to take action to correct any challenges.

3.5 SINGLE SOURCES OF TRUTH

Given the size and complexities of programs there is a lot of information and therefore the management of key program elements becomes critical. Program information is usually stored all over the place including in online documents such as spreadsheets and presentations, project management information systems, in the heads of the team members, buried in meeting notes, and even written on whiteboards in conference rooms. As programs grow in size and as time passes, information gets spread in so many places that it becomes almost impossible to manage. This is why the guiding principle of having "single sources of truth" is so important to managing programs. This is as simple as having one identified source for key information. A key point to note is that these documents need to be constantly updated as information changes, and thus are kept as "living" documents.

There are several benefits to having these single sources of program information:

- There is an authoritative source of program information instead of having information spread across many places. This provides consistency of information across stakeholders and avoids confusion or conflicting information.
- Program stakeholders know exactly where to find key information, which results in better clarity of purpose, scope, and understanding.
- Because these documents are kept current through regular updates, they are always relevant for anyone looking to get the most recent information and can be used for real-time status reporting.

- The program team can respond quickly to risks and problems because the information required to make effective decisions and understand implications is easily accessible and up to date.

Single sources of information can be used as authoritative sources for many program aspects. A few examples are listed in Table 3.1 including some of the important information that can be tracked. By having single inventories such as the ones listed in Table 3.1, a program team will be able to better manage activities and communicate the status of each of these domains. It will also be clear as to how to access key program information.

3.5.1 Techniques

Having master inventories of information means utilizing tools to track and manage the information and having a clear master set of program records.

- *Identify the sources of truth.* It is important to make sure that the sources of information are credible and accurate before collecting the data in one place. It may make sense to inventory all of the data elements and where they come from as a master record of truth for program data inasmuch as most likely these records will come from different systems or files. Program managers may also want to understand stakeholder expectations of where the information comes from and the authoritative sources so there are not differing views on what is a credible source.
- *Leverage existing tools.* If a program is using a project management information system (PMIS) or other control system chances are that it captures some of the information needed. Also most companies have standard tools that they use to roll up to a corporate or portfolio level. These tools should be considered initially to minimize the management of information in several places.
- *Use spreadsheets.* Spreadsheets are extremely effective tools that can be used for managing, organizing, filtering, and sorting large quantities of information. They also allow for pivoting the data quickly in many directions, which is helpful given that different stakeholders need to see information in different ways.
- *Limit access.* It is important to note that not all sources of truth should be shared with all stakeholders. For example, resource lists

TABLE 3.1

Examples of Single Sources of Truth (SSOT) on a Program

SSOT	Description	Information
Program Scope	A master inventory of program scope that lists the specific capabilities requested	• Scope capabilities • Align capabilities to projects • Benefits that capabilities will deliver
Work Intake	A standard "pipeline" list of all requests for work intake into the program	• Description of request • Area requesting the work • Any estimation information • Timing and resource needs • Tracking of where the request stands in the intake process
Project Inventory	A master list of projects within the program	• Project name • A few sentences on the scope of the project • Project manager's name • Budget
Change Controls	A central location for all requested changes to the program	• Details on the change • Impacts on cost and schedule • Disposition as to whether to accept the change
Vendors	An inventory of all vendors, contracts, and invoices	• Vendor information • Contract information such as the name, dates of contract, and cost of the contract • Spend by vendor
Program Financials	A repository for all program financial information	• Forecast by project • Forecast by expense center or department • Actual spend-to-date
Resources	A central roster of all resources on the program	• Resource names • Organizations where resources originate • Capacity and allocations of resources on projects • Role rates used for financial forecasts • Any open roles
Communications	A plan for all communication vehicles and stakeholders	• Meeting information • Communication methods • Stakeholder engagement types
Quality	A central location for all information regarding the quality of the program	• Test case tracking • Defect reporting and progress

that have rates or contract information may be sensitive and require additional access controls.

- *Manage changes.* In order to keep these documents "living" it is important to recognize when information changes. Therefore there should be checklists used during work intake or change control processes to ensure that, once approved, the changes are represented and updated in the master inventories.

- *Identify an owner.* Although these sources of information should be shared across the program, there should be named stewards of these deliverables so that it becomes someone's responsibility to keep them updated and living.

Case Study: Centering the Delivery of a Major Program around a Master Schedule; Government Sector

(Contributed by Nick Pettinelli)

In the world of schedule management, successful delivery is often measured in terms of building a project plan and schedule, completing the work on time, and performing under budget. No matter the truth in these measurements, only the elite programs ever fully achieve complete transparency and total collaboration by creating, managing, and adhering to the concept of a single source of truth. Within many successful programs, the most effective single source of truth (containing comprehensive program information and providing various levels of insight into many management functions) often comes in the form of the integrated master schedule (IMS).

One of the general rules of successful schedule management is to develop a common understanding and agreement of program goals and objectives across the team and its stakeholders. This means that establishing diligent processes, procedures, and structures becomes one of the most critical necessities within any successful delivery. The IMS is the culmination of all vital integrated program structures that are necessary to organize and implement efficient program controls. So, where can key deliverables/milestones, scope, deadline dates, baselines, dependencies, cost/earned value, resources, changes, and risks all be tracked and updated in one cohesive and consolidated single source of truth? The answer is no clearer than the program IMS and the capacity and analysis it is capable of performing. It defines the critical path(s) of the program, captures internal and external dependencies, represents all program scope, and is the primary tool for reporting overall program performance, issues, and change impact assessments. The IMS can provide insight into every functional area within the program; therefore, no other element or management/tracking technique can provide more comprehensive data and act as a better single source of truth than the IMS.

PROGRAM DESCRIPTION AND USE OF A SINGLE SOURCE OF TRUTH

No exceptions were made to this rule when it came time to plan and build the necessary foundation and single source of truth for a large-scale government program. This particular agency was seeking to develop and implement an integrated operating system to support overall field operations. Less than two years' time, an immovable deadline date, an excess of $100 million, and hundreds of resources were dedicated to providing thousands of end users with a highly functional and widely effective system. It was no surprise from the very beginning that the success or failure of this initiative held tightly to a unified source of truth. Team cohesiveness, accuracy, and full understanding of the scope of work was truly imperative to the achievement of deploying the system into production, so no chances were taken on how and where to manage important program information.

The agency and its partners built a proactive schedule management practice, as well as the IMS that accompanied the function, and quickly identified it as the crux to delivering the system successfully, thus earning the program's most coveted and recognizable single source of truth. The program planners set forth on corralling the key functional and delivery leads, and began planning three years' worth of work in less than a two-year timeframe. Key decisions, commitments, and deliverables were the focus of centralizing complex data, and using the IMS as the single source of truth would prove to hold its value in the consistency and dependability that the program leaders and planners knew it would provide.

Firm understanding of the scope of work soon began to evolve into a strategic IMS structure that would ultimately assist in vital decision making. Senior and program leadership, as well as key stakeholders and executives, regularly depended on the information held in the IMS to manage expectations and report on various levels of performance. It contained the program's plan of execution to deliver specified functionality across multiple releases, each of which included key demonstration dates for stakeholder approval. Milestones were logically planned and networked at various hierarchical levels, dates and dependencies were established, schedule risks were identified and managed, and deliveries into production were targeted and agreed upon. The only thing left was to govern strict diligence behind the maintenance of the IMS and provide meaningful status and early predictive indicators that would pinpoint possible troubled areas throughout the program. By nature, the IMS, performing as the single source of truth, was well suited to perform in this capacity.

As are most complex programs, the IMS was eventually challenged as the single source of truth within the schedule management function, but as with all good plans or schedules, it would uphold its title as being ironclad in its dependability. A massive technical change was being mandated to the development of the system, and uncertainty of what would happen to the program's direction was put into question. The change was assessed and an entire re-plan of the release was put forth, but the IMS never skipped a beat. By implementing strict diligence and industry best practices, the schedule team assessed the changes to the affected activities and deliverables, realigned resources and cost, pinpointed potential schedule risks, and re-evaluated the interdependencies and interfaces with other initiatives. The output of this effort was unmatched in its ability to address several moving parts in various functional areas, and report on the results as a single entity. This is how a true single source of truth keeps a program whole, and it's why the appreciation it has from its users is so universally sought after in the industry.

OUTCOME AND SUMMARY

Just as with all typically well-measured programs, this program was claimed a success by deploying the operating system into production on time and under budget. Of course the long and hard hours that team members subjected themselves to was a large contributor to the successful delivery, but what made it truly possible was holding the IMS to the highest standard of the ultimate source of truth. It enabled all levels of leadership effectively to communicate the importance of delivering on key milestones, and it helped the entire team to proactively manage risk and change in a controlled manner. It ensured all types of dependencies were diligently tracked and maintained, it provided crucial insight into the progression of activities and the potential effects to the critical path, and it made certain that there was fundamental transparency into the program's goals. Through these things, the IMS created a positive work environment by celebrating the completion of key milestones, boosted team morale by reminding team members that everyone was working toward a common plan, and instituting trust that leadership was keeping a close watch on all areas of the program.

There were many guiding principles that were deployed to help manage the triumph of this program, but the IMS was the most recognized and widely employed piece of functionality that endorsed successful performance management, and it took center stage day after day as it solidified itself as the most respected single source of truth.

3.6 FACT-BASED DECISIONS

Another important tenet of a consultative approach is being able to use facts to enable timely decisions and actions. There are many decision points on a program related to program direction, scope, issue resolution, risk mitigation, and solution recommendations. Being decisive is the difference between being flexible in responding to events and spinning, which causes delays, low morale, and confusion. The only way to be decisive effectively is to use a fact base to present the case for a particular option or recommendation.

There are many benefits to using a fact-based decision model on programs:

- *Serves as a basis for decisions.* Analysis of an issue, risk, or action item allows the team to identify the options for consideration and consider the implications of each option. For example, resolution of an issue may have a less expensive option, which will fix the problem in the short term or a more expensive option, which is a better solution in the long term. Understanding these implications is important in order to inform the decision makers about the tradeoffs and become a basis for making decisions.
- *Allows for the comparison of options.* Decisions are all about tradeoffs and implications and focusing on the facts allows a program manager to compare different options and determine a recommendation.

- *Takes away blame.* Focusing on the facts of an issue or risk takes away blame and allows the decision makers to focus on what happened and how to respond to it. It also focuses the conversation around the details and not around any one person's political agenda or subjective opinions.
- *Adds credibility.* By leveraging a fact base, a presentation or proposal gains more credibility than an unsubstantiated recommendation. Management is more likely to feel comfortable making an informed decision than a "gut feel" decision, which appears to be based on opinions or speculation.

Here are some considerations when preparing for a fact-based proposal or presentation:

- *Diligence required.* Facilitating fact-based decisions is not easy and does require time and diligence in gathering the facts, performing analysis, and organizing the facts in a way that allows decision makers to understand the options and implications. This is where having single sources of truth for program information can be helpful.
- *Confirm sources.* Calling things "facts" can even be misleading because, in many cases, facts may be a little "squishy" so it is important to make sure that the sources are credible and accurate. This is not easy to do but over time program managers will learn where the trusted sources of information are located. Program managers should also find ways to validate the information by having credible resources review them or finding corroborating information.
- *Determine sufficient analysis.* Sometimes not all information is available (well, most of the time), and there can be the desire to gather and analyze all of it. This can cause more churn and delays. There is such a thing as good enough, and a program manager needs to consider the tradeoff between additional analysis and presenting enough information to guide the conversation directionally.
- *Organize facts.* Just as important as having the facts is the way that they are organized and presented to tell the right story to stakeholders. For example, putting a table of numbers into a presentation does not make it an effective way of facilitating a fact-based decision.

3.6.1 Techniques

There are several techniques to leverage facts for appropriate decision making when running a program:

- *Know your sources.* Based on the last session there should be several single sources of truth on the program that can be leveraged to provide the information needed to enable the facts. It is important to know where key information resides so it can be easily accessed when an issue or action comes up because usually time is at a premium when these scenarios do arise.
- *Utilize QFD.* Quality function deployment is a "method to transform user demands into design quality, to deploy the functions forming quality, and to deploy methods for achieving the design quality into subsystems and component parts, and ultimately to specific elements of the manufacturing process" (Akao 1994). This is a tool where the user lists specific criteria and then has the stakeholders weigh the importance of those criteria. Then each option is ranked as to how well it meets those criteria and multiplied against the weight so that a quantified total can be interpreted. It is a useful tool for comparing options along the same criteria and aligned to the most important priorities of the stakeholders. Figure 3.2 demonstrates this model with an example where the stakeholders want an

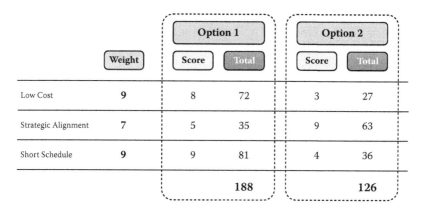

	Weight	Option 1 Score	Option 1 Total	Option 2 Score	Option 2 Total
Low Cost	9	8	72	3	27
Strategic Alignment	7	5	35	9	63
Short Schedule	9	9	81	4	36
			188		126

FIGURE 3.2
Using a QFD approach to compare options.

inexpensive and quick solution more than a strategic solution (based on the weightings), and therefore Option 1 is the best choice (188 is higher than 126). In a scenario where strategic fit is the highest priority (and thus has a higher weighting), Option 2 may be the better choice.

- *Tell a story.* A major part of facilitating a fact-based decision is the way that the story is told. A good consultant is a master storyteller and can craft the information in a way that guides the audience toward a decision. One way is to present the facts and then build up toward the options and recommendations. Another way is to compare different options and show a side-by-side comparison of the details of each option with the implications across several dimensions (e.g., cost, schedule, resources, quality, strategic alignment, etc.).

- *Use benchmarks.* One way to build credibility into a recommendation is to compare data to industry or company benchmarks. This then provides an "outsider view" as a reference point for comparison. For example, if a program is creating estimates it may be useful to compare the breakdown of costs by role or phase as compared to some industry benchmarks to bring credibility to them.

- *Use best professional opinion.* This guiding principle focuses on having facts to make informed decisions; it is not suggesting that every single fact needs to be identified before making a decision. The result of indecision is usually time lost, a problem getting worse, confusion on the team, and possibly missed commitments. Team members need to recognize that they are professionals at their jobs and can use their "best professional opinion" to make a decision given the information that is known at that time and criticality in which a decision is needed.

Case Study: Using Fact-Based Decisions to Grow Insurance Distribution Channels; Insurance Industry

(Contributed by Kevin Savage)

As a national insurance carrier was rolling out new products in the independent agent channel, it was faced with the challenge of building scale among specific agents in target markets to reach a desired customer demographic. The company had invested in building its agency base as part of product launches over a three-year period, however, the carrier wanted to increase its distribution footprint by 50%, adding 2,000 agents to broaden market access and reach more consumers. Doing so required fundamental changes in the process used for selecting, appointing, and authorizing agents to sell affinity program business. Ultimately, staffing, operations, and technology enhancements were needed to enable such growth.

CHALLENGE

To meet the challenge, a program team initiated an end-to-end process review to identify opportunities for improvement, knowing that operational efficiencies would be required to drive faster rates of agency authorization. As part of the evaluation, key stakeholders, process owners, and sales representatives were interviewed and provided feedback based on their experience. The program team realized that although some sales representatives were highly involved with managing agents in the appointment process, many had taken a hands-off approach, leaving it to back-office operations staff to manage issues, creating a less than optimal agent experience.

In addition, much of the feedback on bottlenecks in the process was largely unstructured, had anecdotal input that was not actionable and often created more questions than answers: in essence, it was entirely subjective. As a result, the team initiated a fact-based and data-driven approach to evaluate the agent authorization process in more detail. A sales practice expert along with business analytics resources skilled at turning data into business intelligence were brought to bear on the effort. With the help of these resources, the team identified root causes of issues and found opportunities to eliminate non-value-added tasks, reduce call volumes from agents inquiring on status, deliver enhanced support to field sales teams, and define role clarity for each step in the process. Ultimately, using a fact-based business analytics approach to problem solving, the program team determined there were three key areas of concern:

1. *People.* Role clarity and accountability for each process step was lacking.
2. *Process.* Duplicative tasks unnecessarily elongated the process cycle time.
3. *Technology.* System-generated e-mail messages to agents were confusing.

SOLUTION

People. A team-based support model was implemented in operations and aligned to the field sales structure, creating accountability for every agent in the process and providing key points of contact to sales representatives. Facts were gathered including resource interviews along with a review of sales call history that revealed sales staff were spending four to eight hours per week getting contracts signed and reminding agents to submit required forms. By relieving field sales representatives of these administrative tasks associated with the process, they could focus more of their time on core selling and relationship management activities. In addition, clarifying roles and responsibilities between operations and field sales, assigning ownership for status reporting, and setting clear deadlines enabled all parties to have a better understanding of who was to do what by when.

Process. Within the process itself, the team identified facts for tasks that were inefficient and performed multiple times including obtaining consent forms and contract signatures. The team was able to streamline these duplicative tasks, enabling tighter integration between sales and operations as well as reducing the time for a new agency appointment to complete authorization requirements, going from an average of 60 days to 30 days. The team also established more frequent agent selection campaigns; to be authorized for affinity program business, agents must be selected and invited by sales representatives thus enabling additional opportunities for agents to complete the process.

Technology. A review of agent calls to operations over an eight-week period indicated agency staff were confused about instructions in system-generated messages. To set expectations with agents regarding authorization requirements, the team leveraged marketing communications technology to deliver an e-mail to agents on behalf of their sales representative that outlined the process in three easy steps. This e-mail was distributed in advance of system-generated invites and included an online overview of the authorization process, providing agents a screen-by-screen view of what they would see in the contracting system and directing them to gather required information in advance of starting the process. The team also reviewed and streamlined system-generated messages to agents, making them less technical and easier to understand from the business perspective of an agency principal. Instructions were pared down and focused on the specific actions required of each agent.

Post-process completion, the team built and implemented a new instructor-led 30-minute webinar that provided agents guidance on marketing assets available to advertise and promote their business. The intent of these sessions was to help agents get started with selling affinity business once authorized. Early feedback has been favorable, helping to jumpstart sales activity in the agency channel.

OUTCOMES

Once process improvements were in place, the team evaluated through-put (e.g., the rate of agent authorization) against the baseline it established as part of up-front business analysis. Ultimately, in the span of just three months, authorization completion rates went from 125 agencies per month to 225 agencies per month, an 80% increase. By modifying accountability between sales and operations, the program team enabled greater focus on what each function did best, ultimately generating resource capacity to meet the carrier's goals ahead of schedule. The impact of this effort on business results was significant. By broadening distribution, new business increased dramatically, up 100% as compared to the same period prior year.

BENEFITS

Often times, program teams receive disparate and conflicting information regarding business challenges and how to address them. Sponsors, stakeholders, and subject matter experts all have a perspective of what is working and what is not working as well as input on how operational issues should be resolved to meet desired goals. Although such input is valued, it can sometimes lead to the wrong conclusions and the actions required if not combined with a data-driven, fact-based approach to problem solving that attempts to identify the root cause of an issue and outline solution alternatives. Defining scope based on subjective input as opposed to fact-based intelligence can be the difference between program success and failure.

In some ways, it is not unlike the ladder metaphor referenced by Stephen Covey in *The Seven Habits of Highly Effective People* when describing the difference between management and leadership. That is, "management is efficiency in climbing the ladder of success[;] leadership determines if the ladder is leaning against the right wall" (Covey, 1989, p. 101). In the case of program management, before you start executing, make sure the path you are taking is grounded in data-driven, fact-based analysis that informs business insights on the problem at hand. Reframe the problem as necessary so key stakeholders come to see it in the same light. Ultimately, be sure you are climbing the right ladder; or else you may reach the top only to realize you went the wrong way.

3.7 THE SHIPS IN THE FLEET OF ACCOUNTABILITY

Having accountability means that a program manager feels responsible for everything that happens on his program as well as the outcomes of the program. Accountability, as it relates to managing programs, can be grouped into three buckets that have words all ending in "ship" so I refer to them as the "fleet of accountability" and the program manager is the captain of this fleet.

3.7.1 OwnerSHIP

Having ownership is truly feeling accountable for all of the work, projects, and outcomes of the program. This means not just focusing on a particular set of tasks and hoping others will do their part as well but genuinely owning all work even if a program manager is not directly responsible for doing it on his own.

The benefit of feeling ownership is that it forces a program manager to account for all of the components of the program. Program managers feel ownership of the financials, resources, schedule, quality, and commitments of the program. This forces a different mind-set than if they assumed someone else owned pieces of the program.

3.7.2 StewardSHIP

Stewardship means caring about one's work, the program, and the company. Being a good steward means doing the right things for the company, growing the people in the company, and looking to improve things that are not working as planned. There is no shortage of opportunities to improve processes in companies and organizations.

Today many programs are so large that they have to interact with many different organizations or other programs within a company, which means many different processes and tools as well. Having a mind-set of stewardship means looking for ways to improve those processes or interaction points. It also means helping all of the other programs to deliver more effectively as well. The aphorism, "a rising tide lifts all boats" is a perfect metaphor for stewardship as it can be interpreted to mean that improvements in company processes and practices can improve all projects or programs. This becomes important in larger programs that are often reliant on other projects in the company.

Another aspect of stewardship is the concept of continuous improvement. This means always looking for opportunities to evolve tools, procedures, or processes. This can mean continuing to evolve the program office operations on the program or looking to improve companywide processes.

3.7.3 LeaderSHIP

Regardless of the organizational relationships of the program team, the program manager is the primary leader of the program. Program managers are being held accountable by management to meet commitments and looked to for direction from the project managers and team members. They are also the "face" of the program to all of the stakeholders.

The biggest change needed for program managers in the new business environment is that they must evolve from a program manager who manages the plan with direct control over resources and projects to a program leader who has to use influence and motivate the team through softer skills. Table 3.2 illustrates the differences in these two roles across several program management attributes.

For all of the reasons described in Chapter 1 of this book, the program manager style described above is not effective in the new business environment. Leading programs today require program managers to be able to motivate their team, empower them to be successful, and act as champion for the team.

It is difficult for a program manager to be successful without all three of the ships. Figure 3.3 shows a matrix of these ships and the effects of not

TABLE 3.2

Program Managers versus Program Leaders

Attribute	Program Managers	Project Leaders
Relationship to team members	Have direct reports.	Have motivated followers.
View of role	Team works for them.	Work for the team and act as a champion.
Approach	Manage work.	Lead people.
Decision	Make decisions based on control.	Facilitate decisions based on facts and influence.
Direct	Tell how and when.	Sell what and why.
Power	Uses authority and direct control.	Use influence, passion, and persuasion (e.g., referential power).
Concern	Being right.	Doing what is right (stewardship).
Ownership	Take credit and give blame.	Give credit and take ownership.

Ownership	Stewardship	Leadership	
X	X	X	Full accountability and optimized results for the program and company
	X	X	Oversight and missed work because not on top of all work
X		X	Good for program but inefficient, not continuing to improve
X	X		Inability to lead people and ineffective on program due to morale and poor influence

FIGURE 3.3
Matrix of the "ships."

having all of them. If program managers do not feel ownership of the program, then there will be oversights and missed commitments. Stewardship is what evolves a company's practices, products, and processes and not having it will cause companies not to improve optimally. Without leadership, the team will not be successful or feel motivated to do the work, and key stakeholders will not be influenced on key decisions.

3.7.4 Techniques

Each of the three types of the ships have techniques, which are leveraged for the approach:

- *Ownership.* Define roles and responsibilities. Providing clarity of program roles and responsibilities is one of the most important ways that program managers can increase the productivity of their teams. Programs where there are overlaps or confusion in responsibilities usually result in duplicate work efforts, missed activities, and other inefficiencies. Most delivery methodologies come with a roles and responsibility list (also called a RACI diagram), which outlines key roles and their specific accountabilities. A kickoff meeting should be conducted with the program team early on where clear responsibilities are discussed and documented.
- *Ownership.* We versus They. Having ownership means recognizing that there is one program team and not disparate projects, teams, organizations, or resources. Oftentimes, team members will complain about other people or organizations and say that "they" are causing problems. Program managers need to facilitate a "we" mentality and recognize that everyone is working toward a common program goal, usually to streamline costs or introduce new products that will compete in the marketplace (against the true "they," which is the competition).
- *Stewardship.* Continuous improvement. An important aspect to stewardship is to be constantly striving to improve processes or operations. These can be operational procedures within the program or larger processes within the company. For example, if a program office is tracking vendor information on a spreadsheet, the next logical evolution could be to look into providing value-added services such as contract evaluation or assessment of vendor performance. There are many aspects that can be learned from the Japanese manufacturing concept of *Kaizen* which translates from *kai* ("change") and *zen* ("good"). There are many elements of Kaizen, but the more strategic elements include deciding how to increase the value of the delivery process output to the customer (effectiveness) and how much flexibility is valuable in the process to meet changing needs (Imai 1986).

- *Stewardship.* Sharing is caring. Program managers should also look to share tools, processes, or techniques with other programs or projects in the company. Most large programs have the necessity to create more rigor and tools than smaller programs and therefore usually create robust tools for managing the work, which can be leveraged across the company. Program managers should consider documenting these or making them into templates and then sharing them with their peers.
- *Stewardship.* Grow the talent. One of the biggest stewardship benefits that any manager can do is to grow the talent of the organization. Programs (and companies) are effective because of the talent of the resources on the team. Program managers should recognize that managing talent is part of their job and be constantly looking to improve the skills and acumen of their team members. This could include providing stretch assignments, setting up mentors, offering specific training, and sharing their own experiences and feedback with program team members.
- *Leadership.* Expect and recognize excellence. Foundational to motivating a program team is having clear goals and expectations of excellence for them. By encouraging high, but attainable, goals the program manager is demonstrating confidence in the team's ability to deliver the program and setting the stage for success. The program manager should work with the team to create a project vision statement that outlines these goals. Once the goals have been set, the program manager should look for opportunities to acknowledge and recognize excellence. Team members who feel respected and valued will perform better on their tasks and are more likely to stay loyal to the program.

Case Study: Using the Ships to Succeed in a New Program Delivery Role; Financial Services Industry

(Contributed by Chris Richards)

As a project and program manager for the last 15 years I have learned to understand the importance of the three "ships" and equally how important it is to make them a part of the DNA of how we deliver on our results. In 2012 my role was expanded from a professional development director tasked with career mentorship and guidance to a program delivery director overseeing 50 projects within our infrastructure portfolio.

Although not a stretch given my gravitation toward execution and delivery, it did place me as directly accountable for all aspects of each of the 50 projects within my span of control. There were several advantages to this. I owned the resources and over the past year had worked diligently to build a strong and loyal coalition of high performers. Second, having worked in the division previously as a consultant, I had already built strong ties and relationships with several of our internal technology domain areas and as well the customers I would be supporting.

There were also several disadvantages to my new role. First was the counterbalance to my last point in that I had ownership over it all. As a leader, we have to own the failures and challenges as much as the successes. This role would be a true test of that statement. Equally so, the portfolio of projects that I had been given was full of troubled projects, some very high value initiatives, and a highly vocal customer base.

Entering into my new responsibilities I saw my first call to action as stakeholder engagement. Based on the principle of the ships, a program manager has to be a good steward of the programs and projects for which she is accountable (i.e., stewardship). For me, this started with direct engagement of the stakeholders whom I would be supporting. One unit in particular became my focus on day one. They were considered a small business within my portfolio yet for that small business they played a key role in a very important marketplace for the company. Unfortunately for me, at that time most of their projects were in red and yellow status and the customer was generally unhappy with our ability to make and meet business commitments. Understanding the general troubles, I set up a portfolio review in an effort to identify the key contention points and needs of the business as well as what process levers we could pull to help get their efforts back on track. In advance of this I met with each one of the stakeholders on their team independently. This included the individual project team members such as their developers, technical leads and project managers, as well as the members of their senior leadership team. The effects of those pre-meetings were severalfold. Knowing the troubles I did not want to walk blindly into a large customer meeting with the potential of being broadsided by issues. Meetings like that are seldom productive and normally leave people running on defense (the place no one ultimately wants to be). Also, meeting individually allowed an informal venue to build a personal relationship with each person while at the same time identifying key points to develop a go-forward strategy. A mentor of mine instilled in me at a very early point in my career the value of this type of engagement and approach. Essentially never enter into a meeting without the outcome pre-ordained whenever possible. Having met with each person, although I was unhappy that their projects were in such disarray, I was happy to find that fundamentally the issues were not insurmountable. Actually it was the basics that were out of balance. The first thing they were looking for was the approach of stakeholder engagement. Really they just wanted someone to engage, listen to, and be their advocate for getting the job done. I took ownership of the problem to be accountable for all aspects of their work. Quite frankly, we all have to remember that you can't fake leadership. To quote Lee Iacocca, "Lead, follow, or get out of the way."

I took these leadership principles to heart and gave each of the stakeholders my personal cell phone number. In doing so I also told them to call me day or night regarding any issue or challenge at hand and (trust me) they did. With their program lead I also set up a weekly one-on-one just to keep the lines of communication open. In those meetings he and I spoke frankly about the issues and we strategized on how best to "move the ball down the field" without politics. That individual and I ended up tied at the hip for our work and I consider him as much a good friend as I do a business partner. If there was one catalyst for success I would say it was his incredible willingness to meet in the middle, put everything else aside, and focus on the mutual success for the business. By doing so, we were truly able to innovate and optimize how the work was getting done.

I was also able to identify that our approach as a highly matrixed organization was actually causing problems for their business. In the end, they only ran about five to six projects a year; however, we had four to five project managers working on their projects. Thus communication, governance routines, as well as priority and delivery were not coordinated in any particular fashion. To rectify this, I organized their projects into a program under a single program manager. The person I assigned was one of my best resources, had a long history with this particular business, and was also a strong and excellent relationship manager. Ultimately it was about logical groupings of deliverables and how best to prioritize and control the work delivery. Organizing the projects into a program allowed us to govern effectively both the demand and supply side logistics of the various environments the customer needed. From there we built a true integrated plan that went further than just simple dependencies but also took into account benefits delivery and realization. This gave the team and leadership insight into where compromises existed. Some projects within the program were slowed due to resource constraints and other challenges so other benefits on dependent projects could be realized earlier. In doing so we made a win–win for all.

Over the course of 12 months and working together (customer and delivery team alike) we iterated around our approach. It proved to be fruitful and although I will not say that the road was paved with rose petals, there were many improvements. We worked together as a team to be better stewards of our work and stakeholder engagement, demonstrated leadership in direction and resources, and took ownership of our end-to-end deliverables. I have moved on from my former role; as I wrote this case, however, it is interesting to note that the weekly one-on-one that I had with my former business partner still remains. These days although our conversation has changed, the theme remains the same: how, as partners, we can steer the ships to collective success. In closing I wish you fair winds and following seas on your respective projects and programs.

3.8 SIMPLICITY

When I first started as a programmer after college I learned that there are many different ways to develop specifications and many different coding features that could be used. I would create the most brilliant architectures and technical solutions (according to me, of course) using these cool features. The problem was that they resulted in complexity to support and maintain and probably went well beyond the business requirements.

As I have gained experience in my career I have learned that simpler approaches are almost always better than complicated ones. This applies to the development of application code, documenting processes, or any other solution that I have tried to implement. I have realized that there is a curved relationship between my experience and the complexity of

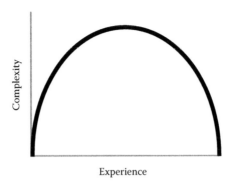

FIGURE 3.4
Relationship of complexity and experience.

the solutions that I have implemented. Figure 3.4 shows this relationship where as I first started out I made things really complex, but then as I gained experience I realized that I did not have to make things complex and learned to simplify them.

Over the years I have seen countless examples where well-intending people have overcomplicated a solution or process. The net result is long lead times and complexity that have built up over time and are hard to unravel. Now I try to make a conscious effort to keep everything I do as simple as possible. Based on the curve in the figure, now when I see something that is simple, I realize that the person who created it either did not know better (left side of curve) or was a genius (right side of curve).

There are many benefits to taking a simplistic approach to work:

- Simple solutions require less maintenance and therefore are less costly to the organization supporting them.
- Simple visuals in presentations are easy for stakeholders to understand the key concepts so that they can interpret information or make key decisions. Important decisions can be delayed if the stakeholders are confused or cannot understand the key points from within too many details.
- Simple processes are better for team members to follow, and they focus on the important concepts and not every granular detail. Complex operational processes or templates will confuse team members and result in inconsistent work and a poor reputation of the program office with the team. Team members can also be overwhelmed

with complicated processes that have many steps and then choose to avoid them altogether.

- There is such a thing as good enough at times. By focusing on keeping things simple and relevant a program team can become much more efficient with its work, which will make it more productive (and valuable).

The concept of simplicity is extremely relevant to programs and especially for operational processes. Program teams should try to make the processes, tools, and procedures as simple as possible for their users. Consider the advice of Antoine de Saint-Exupéry (1939/1967) when he said, "It seems that perfection is achieved not when there is nothing more to add, but when there is nothing more to take away."

Note that this guiding principle may seem at odds with the guiding principle of attention to detail (i.e., simple vs. detailed). Paying attention to details means understanding the details and ensuring they are accurate and high quality, but those details need to be managed and organized in a way that makes them easy to understand. Thus simplicity is not about staying away from the details, but rather it is a mind-set on how to approach and present the work.

3.8.1 Techniques

Keeping things simple is actually not that difficult but requires deliberate planning. There are several techniques for approaching program work this way:

- *Focus on the necessary.* Understand what the requirements are for a particular process, tool, or solution and then anchor them around these requirements. If additional elements are needed to be processed or analyzed they should be handled in a different part from what the stakeholders see so there is not too much clutter. This then allows the stakeholders involved to stay focused on the key points. For example, a presentation on a specific program issue should focus on clearly presenting the impact and options; the analysis and details can be put into the appendix.
- *Automate as much as possible.* By leveraging existing tools or automation, information can be generated instead of manipulated and

derived. A project management information system can be used to gather, automate, and report on program information.

- *Draw pictures.* Many people process information visually and pictures are an effective way to explain complex concepts by showing objects in relation to each other. Obviously the pictures should not be too complicated as that would go against the principle of simplicity and may create more confusion. For example, a presentation on the scope of a program would be better received if the information were organized into buckets and grouped together than just listed in a bunch of bullets.
- *Organize information into categories.* Instead of presenting large quantities of information, look to group it into categories or buckets. One example of this would be to group information into specific stakeholder needs to give them each the information being sought.
- *Try not to customize.* Companies that buy technology solutions get into trouble when they think that they are "different" and therefore try to customize the solution. This almost always results in additional maintenance, unplanned cost, and problems. For example, if a company purchases a customer relationship management (CRM) system to manage customer information, they should not try to then add new customized features related to underwriting and marketing functions which are features that are not naturally done in this type of system.

Case Study: Using Simplicity to Manage a Complex Program; Health Services Industry

(Contributed by Melissa Brickhouse)

In 2011 a program was initiated to provide new health care products into the market to help reduce the overall cost of health care. The program cost $10 million and was an 11-month effort. In order to deliver the program, we had to interact with 30+ departments, and nearly 400 resources. The program deliverables drove end-to-end changes including but not limited to sales, case installation, the company website, and financial reporting. This was a key program, inasmuch as it directly aligned to the strategic imperatives for the company's growth. Failure was not an option.

CHALLENGE

The program faced two challenges: the first was a lack of awareness of the new product, and the second was the ability to make the decision process efficient. The team had to simplify the program to achieve its goals.

APPROACH: AWARENESS

The program introduced a product to the organization that was a new way of doing business. Many people had a hard time grasping the impact of the new product and struggled to translate the idea into how it would affect their applications or operations.

In order for the program to meet its tight implementation timelines, they had a compressed four-week requirements stage, which was shorter than most programs. Requirements meetings were held and a lot of redundancy came up in the Q&A sessions. People in various application areas continued to ask the same question in a slightly different way hoping to understand the changes. Even with the answers provided they still were not clear. To address this problem, the team began thinking of concepts to ensure analysts fully understood the scope. The team had to simplify the product into terms that the application areas understood. The program team did this by getting to common and familiar terms.

The program team also held scenario sessions to walk through the new concept. In the sessions the program team would state how the new product was similar to an existing product, but would also state how the new product was different from it. The methodology used helped people understand the associated impacts. Once the associations were made, the light came on for most of the application teams and they began to change their questions into comments or statements. Instead of asking, "What is this product?" analysts started stating how it would work. Then the analysts focused the questions on the perceived differences for clarification purposes. The amount of new questions for the analysts was substantially reduced. Because this approach was successful, the program team continued to associate the new product with familiar concepts when beginning conversations. The end result was the program's ability to move forward to close the requirements with a strong direction on what changes were needed to make the product vision a reality.

APPROACH: DECISION MAKING

The program looked to provide client reporting as part of the delivery. In doing so they had two very different options from which to choose. Depending on which area the program team talked to, they would get a different decision on which way the reports needed to be generated.

When trying to get to a final decision the program team felt it was like a ping-pong ball in an Olympic match. With every meeting the program team held, they could not get the information needed or the people to move forward. Sometimes the program team would get a decision, but it wouldn't align with the strategy, or wouldn't align with the timelines they had. There appeared to be no middle ground. The program team had to simplify its decision-making process by eliminating the back and forth. This was achieved by following a similar version of the mathematical order of operations, "Please Excuse My Dear Aunt Sally," that is, P, E, M, D, A, and S.

When solving a complex math problem, the order of operations is crucial if you want to come to the right answer. The concept can be applied to simplify program management. When making decisions within the program, we followed an order of operations to get to the appropriate answer.

The first thing the program team did was deal with the P, parentheses. The parentheses were the constraints the team had for the program delivery. The major constraint was time. The program had to get the product to the market in time for the renewal/selling season.

Next the program addressed the E, exponential. For the program, this was the architectural direction. The program team worked with the architectural team to ensure that the solution was in line with the strategy and would handle the anticipated growth.

From there the program focused on M, multiplying, and D, dividing. This was equivalent to the program crunching numbers and doing the required research. The program team reached out to the affected areas and gathered costs and impacts of the solution with alternative designs that would still align with the constraints and architecture. The program team worked with the application areas to provide justification of the costs.

Finally the program team focused on A, adding, and S, subtracting. This was the program weighing the pros and cons of the solution. The team had options to enhance the legacy systems or build out a strategic data store.

The program team provided all parties with the details needed to make a decision prior to the meeting. The meeting was structured so a decision could be made during the allotted time. Through this process the program team learned to simplify getting the answer to a problem through a repeatable process.

SUMMARY

As a result of simplicity in awareness and decision making, the program was able to meet its aggressive commitments even with a very complex program team and structure. Simplicity was something the program team continued to reflect on as they went through the development lifecycle. Whether it was communications, requirements, design, or status reporting, "Are we making this harder than it needs to be?" is a question that the team often asked. The program was able to meet the commitments and deliver the new products in the marketplace. As the program team moves forward with new programs, they continue to look for new ways to simplify our program and maximize our results.

3.9 TAKING A CUSTOMER-FOCUSED APPROACH

The final guiding principle is to take a customer-focused approach to all program work. Most companies today have the "customer" somewhere in their mission, slogan, objectives, theme song, and so on. This is a top priority for companies as expressed by their leadership given that customers buy products and services, recommend companies to friends, post feedback on social media outlets, and directly influence profitability. It is ironic how much importance is placed on customer experience externally for companies but how little is done internally between divisions that have service-customer relationships. That is, many divisions do not treat the internal "customers" of their services the same way that they say they do for "real" customers. Here are some classic examples:

- *Not responding to messages.* This puts the onus on the requestor and customer of the information to follow up to get information. Not only is this poor customer service, but it also demonstrates a lack of ownership.
- *"I do not do that."* I hear this all the time from people in a division, and I need their services. So instead of acting as a liaison for their division they are making the customer determine their internal organization's structure.
- *Not meeting commitments.* After committing to a date or activity the work does not get done, and program managers often find out only after they have asked once the date has passed instead of proactive communication or management of expectations.
- *Poor quality.* Having the customer find the defects or issues. For example, the customer of code may be the testing team, and code that has not been properly unit-tested then becomes the problem for the tester to find.

If a company consistently did any of these things to their external customers, they would be out of business rather quickly. Yet many internal organizations operate this way, and somehow seem to be comfortable with this approach. The difference is that in the external marketplace customers have choices about who they should actively support with their business. We usually end up with the internal organizations that we work with and need to figure out how to partner for success. Table 3.3 highlights some examples of internal "customer" relationships on programs and the services that are offered.

All program team members should look to understand who their customers are, which includes other organizations, other projects, as well as the affected customer of the business or program. Then they should recognize that they would not have jobs without these customers and should view their relationships as important ones and continually focus on them. They should then look to provide as much value as possible to their customers, so that they constantly operate using a "customer-focused" mind-set.

3.9.1 Techniques

Having a customer-focused approach is a mind-set that program managers and all team members should have. This approach enables the team to consider customer needs and view themselves in the eyes of their customers, which results in higher quality and satisfaction. It also improves

TABLE 3.3

Examples of Internal Customer Relationships on Programs

Service Provider	Customer	Description of Service Provided and Value
Program Office	Projects	• Aggregation of program information and insights into progress • Standard processes and tools • Operational services such as contract, financial, and resource management
Program Office	Management	• Management reporting including health dashboards and status • Insights into trends and progress • Information to support the selection and prioritization of programs and projects (e.g., estimates, timelines, and resources)
Sourcing Office	Program Management Team	• Tracking and negotiation of contracts • Performance and service level agreement management of vendors
Finance Organization	Program Management Team	• Accurate reporting of program financial information
Related Projects	Program	• Meeting agreed-upon commitments of scope, cost, quality, and schedule
Resource Managers	Project Managers	• Availability of resources with required skills to meet commitments
Business Analysts	Designers and Developers	• Quality requirements that accurately identify the functions of the program
Business Analysts	Testers	• Clear identification of the functions that need to be tested and the expected results
Developers	Testers	• Quality code that has been initially tested against the requirements
Program Team Members	Stakeholders	• Quality solution that meets the requested, committed to, functionality

morale by having team members feel more aligned with their customers' expectations and perspectives.

- *Focus on value from a customer perspective.* For every activity, deliverables, and interactions consider who the customers are and their interests. It may even make sense to document these preferences as requirements and review them with the customers to confirm expectations.
- *Manage expectations.* It is important to manage the expectations of stakeholders and customers so that everyone is on the same page with regard to assumptions. This can include expectations of when a

program activity will complete, expectations around what scope will be delivered, or expectations around any tradeoffs or implications or decisions. Consider a situation where the program work will take 12 months to complete. If the expectation is set that the work will be done in 14 months and the project comes in two months early, then the program manager is seen as a miracle worker. However, if the expectation is set that the work will be done in 10 months then the project manager comes in looking as if he cannot meet the requirements. However, in both scenarios it took 12 months to do the work. Therefore, having a solid estimate and setting realistic expectations up front is critical.

- *Solicit feedback.* Asking customers for feedback is one of the best ways to understand their interests and opinions on the relationship. This way service providers can understand what is important to their customers and make sure that they can focus on providing the value for these priorities. Checklists or surveys are both effective techniques to solicit customer feedback.
- *Focus on continuous improvement.* A constant focus on improving processes or tools is also important to a customer approach as we should never stop striving to provide more value and service.

Case Study: Taking a Customer-Centered Approach during Site Disposition; Pharmaceuticals Industry

(Contributed by John Moleiro)

A biopharmaceutical company, participating in the development of prescription medicines for humans and animals worldwide, acquired another large pharmaceutical company. This $68 billion acquisition encompassed technology integration for targeted locations across the research and development facilities in North America. The enormous acquisition resulted in a need to swiftly and effectively integrate operations of the two companies. Realizing synergy savings through the reduction of redundant business capacities and sites was one of the objectives. Once the acquisition was official, the IT Global Program Delivery Office formalized a "Site Transitions Program" and identified it as one of the key providers to accomplish synergy savings across all integration programs creating decisions to:

1. Exit sites completely.
2. Consolidate sites with other sites.
3. Reduce existing site footprints.

In combination with the dire timing of the decisions it was vital to maintain the intellectual property assets by:

1. Retaining knowledge during the disposition of data, applications, infrastructure, and people
2. Reallocating information to ensure the legacy data align with business continuity
3. Guaranteeing information retrieval will continually fulfill the need for legacy information in the event of a US Food and Drug Administration (FDA) audit or legal inquiry

To enhance the synergy goals these events had to be conducted with utmost promptness. Taking a customer-centered approach was vital while focusing on site dispositions, particularly when it came to making decisions regarding people movement. The focus on the people during the site dispositions inspired innovated processes to produce an outcome offering the best customer experience humanly possible. Before we go into how this was accomplished, we define who the "customers" are and how they were affected by the site dispositions. Customers were categorized as:

1. The individual who is being moved from Site A to Site B (the end user). It was critical that the end users would maintain productivity once located in the receiving site, maintain security integrity of the data during transportation, and ensure efficient preparations on the sending site for an effective transition.
2. The supporting functional and enabling teams (i.e., data migration, account management, help-desk, messaging, procurement, human resources [HR], etc.). During the transition from sending to receiving sites there are many interdependencies that must collaborate and work closely together.
3. The site transitions team. It was a focused provisional entity that developed, managed, controlled, and measured the overall process of the person's transition from both sites.

TAKING A CUSTOMER-CENTERED APPROACH

With company integrations, after an acquisition, the natural effects on processes and toolsets not being fully rationalized causes diverse interdependencies and redundancies among the functional and enabling teams. There are many scenarios that can be described on approaches to offer a good experience for the end user (mainly), but also the functional and transition teams. For this case study, I focus on two situations that were the more prevalent.

1. Interdependent functional team collaboration breakdown
2. End-user collaboration and expectation breakdown

INTERDEPENDENT FUNCTIONAL TEAM COLLABORATION BREAKDOWN

In a complex organization where end users rely on the toolsets to manage data and maintain efficient productivity, there are clearly many supporting teams that must work cohesively to produce effective results. Now imagine a company with two procurement systems, or two on-boarding systems; image two help-desks, phone systems, and infrastructures constrained with different environments impeding the ability to access data. Considering this complex integration scenario, it would have been easy to have a breakdown of communications among the diverse functional teams. So how did we proactively avoid these types of breakdowns? One of the key aspects was to nurture internal partnerships by understanding all of the functional teams and their respective processes. By having frequent synchronization meetings among each key functional team we maintained the collaboration and were able to make quick adjustments to necessary processes. Daily issue management focus groups helped to mitigate immediate issues as they arose. As an example consider that an end user is coming from the acquired site (sending) to the new site that is managed by the acquiring site (receiving), and there was a problem connecting to vital unstructured or structured data. Does the end user call the help-desk from the sending site or the receiving site? If the help-desks from both sites were not originally recognized to understand each other's processes, then the result would not just have been an unsatisfied user but loss of productivity and possibly additional costs incurred. Now change the scenario around a bit. Even though there were two help-desks, by recognizing and proactively understanding the end users' expectations via a checklist of questions, the risk could have been identified, offering the appropriate time to determine a mitigation plan between both help-desks. This mitigation plan allowed the site transitions team to collaborate with those two functional areas to determine the best temporary process for the time being. This forced not only a better relationship with the end user but also with the functional teams.

END-USER COLLABORATION AND EXPECTATION BREAKDOWN

Large organizations are generally complex, but now we were adding an integration of two large companies with site transitions while people were moving. It was critical to define and understand the end user expectations to offer optimum results. By not understanding the end user's expectations, on-the-ground supporting teams would have been reactive in an attempt to resolve issues causing unsatisfactory service-level agreement (SLA) responses. Many challenges could have occurred if we were not utilizing those tight partnerships among functional and enabling teams. The challenge was that while those functional teams were still dealing with their own integration challenges, we didn't know who was listening to the end users when they were struggling, and resolution was taking too long. One of the methods we used was to leverage the site transitions team. The site transitions team pre-set end-user expectations by operationalizing provisional safety nets such as:

- Focused mailbox and calling response paths so that if the help desk was not able to resolve issues, the site transitions team would intervene and mitigate the issue (e.g., create temporary processes, collaborate with specific teams, build focused teams, etc.)
- Proactive surveys at multiple checkpoints during each end-user move/transitions

One of the more successful approaches was based on the second bullet above. Historically using surveys is most useful in determining how to improve processes after the experience, which can be too late if the problem has already occurred. To resolve this dilemma, collaborating with the end user more often during the move process offered the ability to set correct expectations. By not having a survey at the end of the process by offering the ability of staged visits proactively with the end user, during predetermined process timeframes (prior, during, and after to include multiple visits even after the process completion) increased the ability to mitigate issues more successfully. During the visits, questions were asked focusing on:

- Any existing issues at hand
- Feedback to improve the process to date (lessons learned)
- Success stories to share to date (what was helpful during the process)

Shortening the timeframes between visits and collection of these datapoints proactively allowed for a balanced approach not only to set expectations with the internal customer but also to have the ability to reduce the timeframe of the issue while collecting improvements more quickly for immediate mitigations. And, of course, we needed to maintain the relationships of the functional team so that if new issues arose then immediately to react and resolve them effectively.

BENEFITS TO TAKING A CUSTOMER-CENTERED APPROACH

Having a program delivery office surely benefited the ability for a successful integration as a whole focusing on a managed approach for the program. This created a holistic approach to manage scope, analysis, planning, and execution of this complex program by:

- Driving the program at a strategic level and frequently reviewing the program performance to ensure the fulfillment of the business objectives
- Applying program methodologies for planning, execution, reporting, watching the health, and delivering on the objectives of the site transitions
- Mitigating risks, issues, and developing workarounds
- Managing interdependencies with other programs across the company
- Negotiating and resolving conflicts among processes, people, and toolsets

In addition to the program delivery office overseeing the entire integration, by having a site integration team in place, specifically around people moves as one of the major activities, also benefited both the functional supporting teams and most important, the end user. The site transitions team interjected a cohesiveness to mitigate, coordinate, and manage all aspects of the people-move process. The partnering, functional, and enabling teams had the ability to leverage the site transition teams for resource supplementation, process creation and improvement, and increased collaboration synergies. As for the end users, they benefited by having a safety net specifically during the people-move process. The benefit of that safety net allowed the ability for the end user to reach out to the site transitions team at any time in the process in the case that the business-as-usual teams were not able to perform their functions due to the natural gaps of the organization's integration efforts. The end user also benefited by having the ability for proactive surveys being performed during predefined stage gates during the transition processes for real-time feedback. This real-time feedback enabled quicker turnaround for issue mitigation and proactive improvements to processes. Real-time feedback also fed into more useful data for lessons learned and success stories during the transitions and not waiting until either process completion or overall integration program closure.

CONCLUSION

In conclusion, the most important aspect to a successful transition when focusing on people moves is that:

1. Data integrity remains intact.
2. Tool sets/application remain functioning post migration.
3. Downtime is minimized during the migration process.

If any of the above dynamics is not satisfied, the end user is now at risk. Clearly the ability of surveys can be used to gauge the person's disposition experience, but the challenge is that if anything went wrong and was stated in the survey as such, then it is too late to act. The goal is not to have any issues occur during the process, but then again, perfection is not a realistic goal either. By having a site transitions team acting as the "safety net" to the transition process of people moves proactively mitigates process and tool gaps during the integration of the two organizations. Despite the processes and the tools used to manage the end-user site dispositions, not presetting the internal customers' expectations in parallel with the supporting team relationships can cause a poor service experience. Even with the ability to utilize the functional principles of managing programs to maintain within the triple constraints of cost, schedules, and the quality of the services, the process could still be at risk. In fact, we may have to add another element to the triple constraint and call it the "quadruple constraint." The fourth constraint should be the focus on the "customer experience" where the functioning principles of program management are just not enough to maintain the ability of internal customer-service relationships.

3.10 USING THE PRINCIPLES

Although each of the guiding principles was described individually in this chapter, it is important to know that it is the combination of all of them which produces successful outcomes. For example, a team that has diligence but not transparency means that team members and stakeholders will not be aware of the hard work being done or how they are progressing. Also a team can have single sources of truth, but if they are not set up to be customer focused then the information may not be valuable or insightful to the program stakeholders.

This chapter concludes with a case study that demonstrates how using all of the guiding principles had a successful outcome on a program and the next chapter goes into details of how to apply the principles to the different functions of a program.

Case Study: Using All the Guiding Principles for a Successful Site Move; Financial Services Industry

(Contributed by Amy Cordova)

This case study summarizes a site operations and technology move from an existing location to a new location (new physical building) where the entire scope of planning and execution of multiple iterations of moves was completed in less than 12 months. Planning occurred over a period of six months and the execution period occurred in weekly iterations over a six-month period. The moves included approximately 2,000 customer service representatives and associated administrative staff, corresponding data center infrastructure, network, telephony, and a wide range of technologies and software applications from mainframe terminal emulation to industry-leading financial software packages. The program team consisted of approximately 40 technology team members with focused skill and expertise in the corresponding technology areas: network operations, telephony, software, and so on.

PROBLEM STATEMENT

For this particular program, the problem to solve was completing all iterations of moves with no downtime. The customer service representatives for this company support callers on a 24/7 basis, so downtime or site outages were not allowed. As such, the planning for the move was critical to ensure that no downtime resulted. A considering factor with this program was that the company was moving to a newly constructed building. This provided some opportunity for preparation and testing, however, this also provided challenges when technology and infrastructure did not work as expected. This required creativity and robust issue resolution processes to meet the requirements for each move.

APPROACH

First and foremost, the entire team operated with a customer approach in mind. In other words, the guiding principle of the program was to avoid impact on the team's internal customers (the customer service representatives). The team leveraged the guiding principles described in this book in the following ways:

- *Leadership.* The team set up a very clear structure outlining the leadership team and which leader owned which function (such as desktop computers, network connectivity, voice, etc.) and the provided transparency across the entire program. Leaders collaborated well together, made sure all communications to their teams were clear, and they exhibited continuous support and recognition for their teams.
- *Transparency.* All information was available to the entire team. This was before the days of tools such as Microsoft Sharepoint, but the team knew that collaboration was critical so they set up

a shared folder structure with a common taxonomy for supporting documentation, and all documents were shareable. This was critical to maintain because desktop PC information had to link to network ports which had to link to local area network (LAN) closet locations which had to link to server information and so on.

- *Diligence.* One of the most critical things done was to assign a particular department or floor move to one of the team members. This provided ownership for ensuring the move was completed accurately and on time. These assignments empowered the lead team member who then remained diligent in terms of planning and executing his portion of the move. It was a bit of healthy competition of sorts because no lead wanted his section of the move to go poorly. Best practices and lessons learned were shared across the team for improvement with each move.

- *Attention to detail.* Attention to details was a "must have" for this team. If any of the data elements were incorrect, then getting a workstation and a customer service representative up and running was delayed while the root cause of the problem was traced back to the source. As such, attention to detail was ingrained in the team as a top priority and data were always verified and re-verified before changing any of the shared documentation.

- *Single sources of truth.* Given the fact that the program was dependent upon a high volume of shared data, having a single source of truth for information was critical. This was an interesting challenge because of the critical need for transparency, so the team needed a mechanism whereby a single source was used, but multiple people could maintain and update through the program's life cycle. As such, rules had to be established to ensure appropriate nomenclature, change control, and usage.

- *Fact-based decisions.* With this program, to avoid downtime, all physical moves were completed in "waves" over a six-month period. The team completed the necessary moves and connectivity updates during lower call volume hours, and had to coordinate very closely with the business leadership to ensure the correct priority and execution of the moves. The team often had to make quick decisions on any changes requested, so committing to fact-based decisions was critical for resolving issues as they surfaced.

- *Simplicity.* Completing a move like this is not highly complex from a customer service representative perspective as the end user. The end user leaves work one day at one location and shows up at the new location the next day expecting to see all of the workstation needs such as a computer, phone, papers, and personal items. Although this seems simple, there is significant complexity behind some components of the move. As a way to simplify, the team focused on the technology moves set up and tested as much of the connectivity as possible ahead of the move date. A task force was made available for every move for immediate troubleshooting. Processes were set up and lessons learned applied after each wave of moves to ensure the most cohesive approach with each move.

- *Taking a customer approach.* With this program, the guiding principle was no downtime for the customers. The entire team operated with a customer approach in mind so there was a strong partnership across the business and IT throughout the duration of the program. Team members frequently asked, "What would our customer service representatives think about this." In fact, the team was recognized and awarded by many of the business stakeholders for their efforts and their ability to operate with the customer in mind.

The application of these principles was as important as the principles themselves. Communication was the most critical component both in terms of planning and setting forth a communication plan to be used for the six-month execution/move period and throughout the moves themselves. In addition, diligent program operations were critical in terms of the workforce logistics and document management as both needed to be very clear and maintained in a consistent manner. Schedule management was also critical due to the significant number of dependencies and milestones to track with every sequenced move. Lastly, the team was responsible for managing vendors, change management, issue and risk management, and of course decision management in the fast-paced iterations of moves that were completed.

OUTCOMES

The outcome achieved by this team was moving 100% of the customer service representatives, desktop computers, and associated infrastructure into a new building with no disruption of service, no downtime, and a move that was essentially invisible to customers. A physical move is generally not complex, but when combined with new infrastructure at a new building, there were definitely things that "went wrong" and problems that surfaced. With the team adhering to guiding principles, they were able to complete the work efficiently and address and resolve problems efficiently and effectively.

SUMMARY

By remaining vigilant on the guiding principles and by having a team employed and driven by a common understanding, the team was able to achieve success and move this customer service center with zero disruption. Teams need to tailor how they follow these principles, and there are a number of techniques to use and apply these principles. With a strong vision and a team able to communicate and execute, the application of the principles for what the program needs is a valuable focus that will benefit the team throughout the program's life cycle.

4

Program Management Functions

4.1 OVERVIEW

In 2011, Project Management (PM) Solutions conducted a survey to identify the top causes of information technology project failure. The report, called, "Strategies for Project Recovery," covers 163 companies of different sizes, which manage on average $200 million in projects each year. The study identifies the five top causes of troubled projects (PM Solutions 2011):

1. *Requirements.* Unclear, lack of agreement, lack of priority, contradictory, ambiguous, imprecise
2. *Resources.* Lack of resources, resource conflicts, turnover of key resources, poor planning
3. *Schedules.* Too tight, unrealistic, overly optimistic
4. *Planning.* Based on insufficient data, missing items, insufficient details, poor estimates
5. *Risks.* Unidentified or assumed, not managed

Tables 4.1–4.5 break down each of these failure points and show their alignment with the guiding principles of a consultative approach, which should be used to mitigate these risks or would have avoided these failures on programs altogether.

Based on these tables, it is clear that using the guiding principles of a consultative approach will directly affect, mitigate, and even avoid these main causes of failure on projects and programs. These guiding principles need to be applied to all the different functions within a program in order

TABLE 4.1

Alignment of Failures to Guiding Principles: Requirements

Guiding Principle	Description of Mitigation and Impact
Diligence	Having the diligence to capture all program requirements, ensure they are consistent, and get signoffs to confirm agreement and understanding
Attention to Detail	Ensuring that the detailed requirements are accurate, prioritized, precise, not contradictory, and clear
Transparency	Ensuring that the requirements are clear to all team members to enable agreement
Single Source of Truth	A single location for requirements and approvals to avoid confusion
Fact-Based Decisions	An accurate set of requirements becomes the fact base for program scope and all future decisions on it
Ownership	Taking accountability to ensure that requirements are clear and agreed upon
Simplicity	A logical organization of requirements so they are clear
Customer Focus	Having the customers of the requirements prioritize them and agree to them before moving too far into the program

TABLE 4.2

Alignment of Failures to Guiding Principles: Resources

Guiding Principle	Description of Mitigation and Impact
Diligence	Planning around availability of resources, required skillsets, having them ready when needed, and understanding unfilled needs and upcoming demands. This planning avoids conflicts and ensures the right resources are ready when needed.
Attention to Detail	Paying attention to the specifics of resource skills, contract end dates, possible conflicts, and needs for upcoming work.
Transparency	Understanding and communicating resource needs in time to allow for proper planning.
Single Source of Truth	Master inventories including a roster of resources, capacity reports, and open roles.
Fact-Based Decisions	Utilizing resource systems to understand key resource information needed for planning.
Leadership	Effective leadership and high morale resulting in a reduction in turnover of resources.
Simplicity	An easy way to view and organize resource information and understand needs.
Customer Focus	Understanding the skills required to meet the customer needs and then ensuring that they are available and used properly.

TABLE 4.3

Alignment of Failures to Guiding Principles: Schedule

Guiding Principle	Description of Mitigation and Impact
Diligence	Proper planning of the schedule milestones and dependencies to create a practical plan
Attention to Detail	Understanding the dependencies between tasks and managing the details of upcoming activities including confirmation that activities with past dates are completed
Transparency	Tracking the ability to meet the schedule through granular milestones and progress toward them resulting in the ability to take action earlier based on trends
Single Source of Truth	A master schedule that has all program deliverables, projects, key dependencies, and milestones
Fact-Based Decisions	Determining duration based on specific activities to have a practical plan
Ownership	Having accountability for the schedule by setting realistic expectations for when activities will complete and then delivering on those promises
Simplicity	Having a master "one page" view of the schedule that shows the major milestones and communicating it so the entire team is aware of what is to be done
Customer Focus	Managing expectations on a realistic schedule and understanding the critical path to meeting commitments

TABLE 4.4

Alignment of Failures to Guiding Principles: Planning

Guiding Principle	Description of Mitigation and Impact
Diligence	Diligence across all aspects of planning of the program resulting in a higher probability of success
Attention to Detail	Not overlooking any items by managing the details and keeping updated action item logs, issues lists, risk logs, and project schedules and rolling them up into program-level deliverables
Transparency	Tracking progress toward meeting program commitments, which allows time to take corrective actions if needed
Single Source of Truth	Having master inventories of all planning information including the schedule, scope, costs, benefits, action items, issues, and risks
Fact-Based Decisions	Using the appropriate data to plan a program, which could come from control systems or other program repositories
Ownership	Taking ownership of the planning functions to ensure proper plans, accurate estimates, and realistic schedules
Simplicity	Organizing the planning work in such a way as to make the information clear to the entire program team
Customer Focus	Focusing on value-added planning functions, which maximize the chances of meeting program commitments

TABLE 4.5

Alignment of Failures to Guiding Principles: Risks

Guiding Principle	Description of Mitigation and Impact
Diligence	Persistence in identifying risks and creating action plans to mitigate them, which will result in risk avoidance, mitigation, or management
Attention to Detail	Paying attention to the risks and when they are realized and then managing the specific actions to close or mitigate them
Transparency	Monitoring and reporting of risks to inform stakeholders and enable decisions
Single Source of Truth	Having a single risk repository that is reviewed regularly
Fact-Based Decisions	Using data as much as possible to identify or quantify risks as well as to present options and recommendations to resolve them
Stewardship	Identifying risks for the program or company as a way to preserve the integrity of the program work
Simplicity	Not over-complicating the risk tracking process but rather having clear descriptions of the risks and their impacts
Customer Focus	Considering the customer perspective when identifying risks such as determining how they would be affected

to maximize the program's effectiveness. These key program functions are displayed in Figure 4.1, which is a more detailed breakdown of the high-level functions shown in Figure 2.3.

Following are the main functions that a program should execute with a brief description of each:

- *Work intake.* This is the central function for bringing in new requests for projects, resources, or estimates.
- *Schedule management.* Creating a schedule of key activities, deliverables, and milestones with commitment dates. This also includes dependencies within the program milestones and with external program milestones.
- *Financial management.* Managing the budget and forecast of the costs required to run the program.
- *Resource and capacity management.* Understanding the resource needs and availability to meet the milestones on the program schedule. This can include using contractors and vendors to supplement key roles or provide additional resource capacity.
- *Vendor management.* Managing the many vendor contracts and invoices that are used on a program.

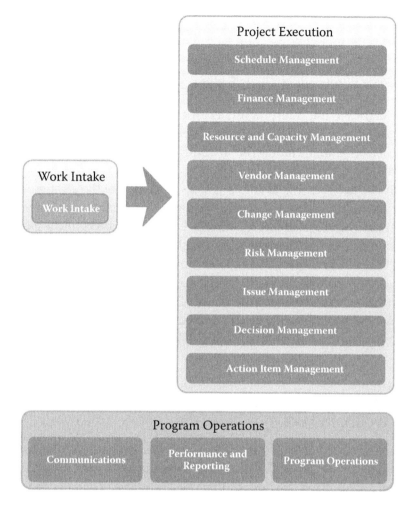

FIGURE 4.1
Primary program functions.

- *Change management.* Identifying any changes from the original program scope, schedule, or cost and quantifying the effects of changes so decisions can be made regarding those changes.
- *Risk management.* Identifying and managing program risks as well as understanding the effects of realizing the risks.
- *Issue management.* Managing issues as they arise, understanding the effects on the program and driving them to resolution quickly.
- *Decision management.* Documenting any key project decisions that are made.

- *Action item management.* Tracking all open actions not related to specific project or program schedule deliverables.
- *Communications.* Standard program communications such as status reporting and program meetings.
- *Performance and reporting.* Developing reporting and performance metrics to track the progress against program and project goals.
- *Program operations.* Managing the different operational components of a program including on-boarding, seating, and equipment.

This chapter goes into detail on how to apply each of the eight guiding principles to the different aspects of running a program to maximize the effectiveness and probability for success. Each section dives into a specific program function to describe it (at a high level because the assumption is that the reader is grounded in these fundamentals) and then describes how to apply each of the guiding principles of the consultative approach. Note that quality management is not identified as a specific function because the premise is that quality is used across all of the functions and is a key component of the guiding principles.

4.2 WORK INTAKE

Because of the complexity of programs and the number of activities that need to be managed within them, program managers should consider setting up a centralized work intake function. This intake function serves as a one-stop-shop for any requests for work, estimates, or resources from the program. There are several key functions within work intake, which include the following activities:

- *Work pipeline.* This is an inventory of all requests for work into the program or portfolio, which could include projects or resources to support dependent projects.
- *Estimation of work.* This can be an initial "sizing" of work to determine the relative cost of the effort and if it is worth pursuing or getting a more detailed estimate.
- *Alignment of request to business and technology strategies and roadmaps.* It is important to understand how new requests align with or affect roadmaps so decisions can be made as to whether these are the

right solutions or activities to work on with limited resources. This is also known as program governance.

- *Resource planning.* Having a central intake function allows for formal tracking of any requests for resources and the determination of available resource capacity to meet those needs. Resources can then be assigned to the work and committed for a start date of that work.
- *Prioritization of work.* If there are competing demands for resources then a work intake function can help to confirm resources are being used on the highest priority items. The prioritization may result in canceling or delaying existing projects but that may be a better use of company resources.
- *Standard initiation.* A central work intake function can also ensure that all new projects in the program utilize the same initiation activities. This could include a standard project schedule and a start-up checklist of activities for the program such as adding project information to single sources and updating program management documents.

An example of a basic work intake process is shown in Figure 4.2. It shows the five primary steps in the process.

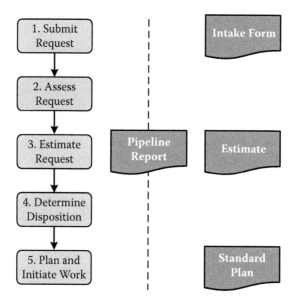

FIGURE 4.2
Work intake functions.

1. *Submit request.* This is the formal step to receive a request for work from a customer, related project, or other stakeholder. Usually an intake form is used, which captures key information such as requestor information, a description of the request, desired time-frame for delivery, confirmation of available funding, alignment to organizational goals, and expected benefits.

2. *Assess request.* Once a request comes in there is usually an assessment activity to understand the details of the request so that the most accurate estimate given the available information can be performed. This can include confirmation of available funding should the project be approved. It can also include a governance function to confirm that the requested solution does not violate any strategic technology or business roadmap decisions. For example, if there are standard technology products being used a new request should not suggest a solution that uses a different technology request.

3. *Estimate request.* After a request is analyzed for information, the program team can then estimate the required work. Usually there are different levels of estimates: a high-level estimate to determine relative size and then a more detailed estimate that identifies specific resources and costs. Assumptions for all estimates usually are documented.

4. *Determine disposition.* A determination of the next steps with the request is then made based on the estimates, prioritization of work, alignment with strategy, and the availability of resources and funds. A request could be denied, approved to start immediately, or approved to start at a later time.

5. *Plan and initiate work.* If a project is approved to start then the intake function should have standard processes to start the planning for that work. This can include a standard project schedule or checklist of activities such as updating master inventories of information, adding the project to the program repository, and so on.

The benefits of having a work intake function are many and include having visibility into all new requests for work as well as a standard way to assess, estimate, prioritize, and initiate requests. Because all work requests will flow through this process, it becomes important to follow the guiding principles within this function.

4.2.1 Diligence

Because work intake is the first step of the project delivery life cycle, it is important to be diligent up front. This sets the tone for the project and ensures that the work gets set up properly for diligent project execution. Diligence during work intake means making sure that all of the activities are tracked and managed properly throughout the process. This includes the following activities:

- *Tracking of all requests.* This entails tracking work intake through the pipeline including an understanding of the work, what step the request is in for the intake process, and the disposition as to how and when to manage the work. Programs may want to consider using a tool to manage the requests in a central location.
- *Proper assessment of the request.* This requires understanding of all the components that need to be estimated. Not being thorough in this step could result in missed costs or assumptions that could result in an unexpected increase in cost or schedule or a decrease in benefits.
- *Sufficient analysis of the request.* It is important to confirm alignment with business and technology strategies and roadmaps. Without this proper governance function, the program could be adding to technology complexity and cost or introducing business solutions that do not align with the strategic direction of the company.
- *Appropriate resource planning.* This is important to confirm that the project resources have the necessary skills required and are also available when needed. Without a resource plan, a project could get the approval to start and then not have the ability to meet commitments because of resource constraints.
- *Thorough initiation planning.* This is important given all of the activities required to manage a program and that need to be kept updated. Initiation planning can include having a standard project schedule template with critical deliverables, updating program information with the approved projects, creating the project charter, and setting up the project structure and templates.

Having diligence across the work intake function sets up the projects for success by ensuring that the analysis of requests are accurate, the project is estimated and resources are assigned properly and that the project starts off in a way that it is set up to be successful. It is much better to be diligent up front than to be reacting to issues, risks, unplanned work, and incorrect assumptions during the execution of the project.

4.2.2 Attention to Detail

Just as with diligence, paying attention to detail during work intake is also important. There are many details that will become critical during the delivery of the project and, therefore, the intake work needs to be as accurate as possible. Some examples of important aspects to focus on include the following:

- *Submission accuracy.* The form used to receive the work request needs to have all of the key elements needed to assess and estimate the work. These details become important to ensure that all of the considerations are evaluated and understood. By not forcing these questions into the intake process they may not be answered, which would result in oversights or unplanned work. For example, the form could ask if there are any infrastructure needs, business process changes, or special testing requirements.
- *Estimation completeness.* It is critical to identify and estimate the key components of the request. This means understanding the solution being requested and ensuring that key affected stakeholders participate in the analysis and estimation activities. Company structures and technology solutions today are so complex that most activities require resources from many different organizations and, therefore, it becomes important to make sure that the key stakeholders actively participate in documenting the work and assumptions.
- *Current pipeline information.* Because work intake is the single funnel of all requests into the program, the pipeline of work needs to be as accurate and up to date as possible. This includes tracking the work requests, updating them when estimates or dispositions are made, and then tracking them into initiation.
- *Accurate initiation checklist.* Most programs have many master inventories of information and, therefore, any approved new requests need to be included in these inventories. Examples could be the master project list, master program schedule, vendor inventory, and resource roster. Initiation checklists should identify these master

inventories and track that they each get updated with the appropriate request information. The initiation checklist should also include other start-up information such as creating the project schedule (and maybe even have a standard template to use), updating status templates, and creating a project file structure. This will allow for the consistent start of projects and give the projects the structure they need to be integrated with the program quickly.

Given that the project expectations will be driven by the initial estimates of cost and time, paying attention to the details is an important approach to setting up the project for success. This guiding principle also allows for accurate tracking of work from inception to start-up and sets up the project well during execution.

4.2.3 Transparency

Depending on the size of the program, there may be many projects and requests for work at various stages of the intake process at any given point in time. Transparency of these intake requests is important to keep stakeholders informed and to communicate what the pipeline of work looks like. There are several ways to provide transparency during work intake:

- *Pipeline report.* Having a centralized inventory of all requests and making stakeholders aware of it provides transparency of where all the requests are in the work intake pipeline as well as key information such as the estimated amount, alignment to strategies, and disposition of how to manage the work.
- *Estimates with assumptions.* Estimates are extremely important to projects because they outline the initial assumptions for the specific work needed to deliver the requested solution. As projects evolve, these estimating assumptions may change so it is important to have clear documentation of the estimates, the components required of the solution, the resources required to deliver the solution, and any assumptions made. The estimation templates should provide for all of this information to make sure the documentation is completed because as time goes on team members often forget the nuances of the estimates unless they are documented clearly.
- *Resource planning.* Having transparency in work intake also includes understanding what resources are required for the work request.

Because most organizations today have resource constraints and people working on many activities, it is important for program teams to understand what resources are needed and with which skillsets.

• *Initiation activities.* There should be a very clear list of activities that a new project coming into the program should have so that all program master inventories get updated and standard templates get used. This provides transparency for the project managers to know what information needs to be updated when they start their project.

Inasmuch as the program work intake function is the first step in the project life cycle and the central place for all requests, it is important to have transparency into what requests have come through and where they are in the process. It is also critical to have transparent estimates with assumptions because these drive the plans and expectations of stakeholders for the duration of the project execution.

4.2.4 Single Sources of Truth

Having a single source of information for all work requests is important. There should be one master inventory of work (e.g., pipeline report) that tracks the request information, estimation amounts, status of the request in the work intake process, and disposition of each item. It is important to keep this information updated as work comes through the initial intake and goes through each of the stages of assessment and estimation. Keeping the information updated will provide confidence and credibility that the pipeline report is accurate and current. The program manager should share the pipeline report and make it available to stakeholders to understand where the work is at any point in time.

Another master inventory that should be stewarded by the work intake function would be to capture the information on the project estimates. Inasmuch as the initial estimates are performed during this function, it would be beneficial to have a library of estimates with all documented assumptions for historical purposes. Many times on projects, initial estimates get revisited or confirmed therefore having a single place for team members to go to locate them would be valuable for the projects.

Lastly, the initiation checklist should be the central source for information on how to start a new project within the program. This should identify all program master inventories that need to be updated, provide standard startup activities, and reference any standard deliverables that should be used.

4.2.5 Fact-Based Decisions

The utilization of a fact base is relevant for the work intake function and especially in regard to the assessment of work requests to determine estimates. Because initial estimates are used to determine feasibility of the request, manage expectations of stakeholders, identify high-level resource needs, and make funding decisions, it becomes important to have them be as accurate as possible. There are several facts that can be used to assess and estimate each request properly:

- Specific business functions required to meet the request as these functions will identify which systems and components of those systems are affected. For example, a business stakeholder wants customer address information so this tells the technology team they need customer information from relevant systems.
- The number of technology systems and components affected by the request as well as an indicator of their complexity.
- Any specific hardware or software that is needed.
- Specialized skills of resources required to perform the work. This can include domain expertise in an area such as claims for an insurance company or technical expertise such as database or rules technologies. The specialized skills may also require the use of an outside vendor if they are not available in the company.
- Any additional expenses that may be incurred on the project such as travel or a specialized product.

All of the items listed above are examples of facts and information that can be used and documented to create an accurate estimate of the requested work. Many times not all of these facts are known up front, but it is important to gather as many as are available and then to make assumptions (and document them) for any additional items. These fact-based estimates then drive key decisions on approval to fund the work, resource needs, and key purchases that will need to be made.

4.2.6 The Ships

The ships have a role to play in the work intake function in several different ways because this guiding principle enables a program team to have a different mind-set of how the intake function should be operating.

- *Ownership.* The intake function should not just monitor people filling out request forms but should seek to understand the purpose of the request and how it affects the different areas of the program. Seeking to understand the intent of the request is the difference between owning the intake requests and simply administering them. This will result in a change in perception of the value of the function.
- *Stewardship.* The intake function has an important role in stewarding the company's resources and money. This means putting the right governance in place to ensure that the requests get vetted against strategic roadmaps and also get prioritized against the other requests to make sure that the program is working on the right activities that align with the company's objectives.
- *Leadership.* In this case, leadership means being at the forefront of the work and providing guidance and direction regarding what work gets performed in the program as well as making sure that the work starts off with the best chances for success. Leadership may also mean having to make difficult decisions regarding the priority of work if one request is more important to the company than some existing work.

4.2.7 Simplicity

Work intake can be a complicated process in large organizations or on large programs so it is important to always consider simplicity in processes, templates, and in the intake approach. This can include the following examples:

- The intake request form should gather enough information to allow for strong analysis and estimation but should look to obtain that information automatically if it already exists in other systems. The requestor should not have to enter information if it already exists somewhere else (this also focuses on the guiding principle of taking a customer approach and making the request process easy for the stakeholders to use).

- The work intake process should also be clear and easy to understand for all stakeholders. There should be a one-page view of the entire process to help manage expectations of requestors and as "marketing" material for the function to use to explain to new team members. This view should include the scenarios for work coming into the process, the key process steps, and then the key output of the process.
- The pipeline report is a central repository of all requests so it also needs to be simple to obtain and understand inasmuch as it will be used often and by many different stakeholders. Users should have the ability to organize and sort by functions that they are interested in such as the requestor name, disposition, or the area that is requesting work.

It is easy for the work intake function to become complicated quickly, but because this is the first time that many stakeholders interact with the program, it will result in the first impression for many people.

4.2.8 Taking a Customer Approach

As with all program functions, the work intake function must operate with a customer-centric approach. This means that the team running this function needs to consider who their customers are and operate in a way that provides the most value to them. In this case, there are a few different customers:

- *Requestors of work.* These are the stakeholders who are asking for program work to be performed including business partners or other projects. For these customers it is important to manage their expectations around the intake process, what information is needed, and how long it will take.
- *Estimators.* These stakeholders are customers of an easy process to facilitate and capture estimation information.
- *Project managers.* These are customers of the estimates and initiation work, and it is important to provide them with accurate initial estimates and a solid foundation of initiation work to start their projects.
- *Program manager.* This is the customer of the information to see all of the work in the intake process including the projects, dispositions, and status of projects in the pipeline.

- *Program stakeholders.* These are customers of information who need an easy way to find information on requests from the pipeline report. This can include sponsors who need to understand when a project will be approved and start.

In each case of the customers listed above, there are different needs that should be recognized and considered within the intake processes and tools. It is the program manager's responsibility to make sure that these customer needs are satisfied and that all customers view this as a valuable function to the program.

4.3 SCHEDULE MANAGEMENT

After a new project is initiated through the work intake process the project manager must define the detailed activities for the project in the project schedule. The first step is decomposing the work into logical components or work packages by preparing a work breakdown structure (WBS), which outlines what work has to get done into work packages. These work packages are then organized into lists of activities that get sequenced and estimated for duration, effort, and resources. Finally, the schedule is developed based on the activity list, activity durations, dependencies, and resource needs.

The project schedule is probably the most critical tool that a project manager has to use because it identifies the body of work required to complete the project and the resources that are performing the work, and is also used to manage and track progress against deliverables and milestones. Figure 4.3 shows the key functions of a project schedule which are broken into two primary activities: create the schedule and manage the schedule.

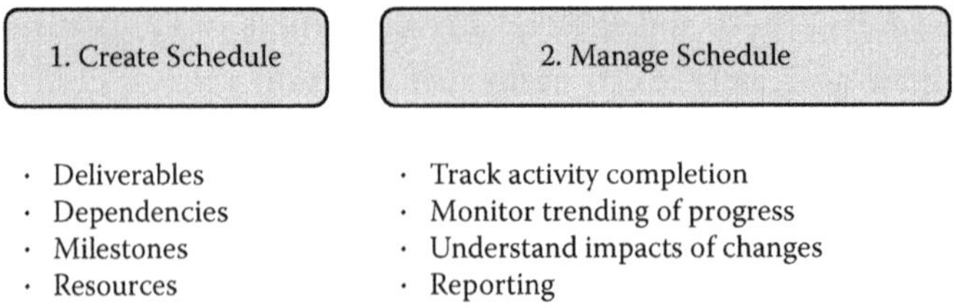

1. Create Schedule	2. Manage Schedule
· Deliverables	· Track activity completion
· Dependencies	· Monitor trending of progress
· Milestones	· Understand impacts of changes
· Resources	· Reporting

FIGURE 4.3
Schedule functions.

The bulk of this chapter focuses on a project schedule within a program but note that the guiding principles are applied the same for a master program schedule as well.

4.3.1 Create the Schedule

In this initial step, the project manager identifies the work that needs to be performed and documents the activities, deliverables, and milestones that are required to support that work. Usually these activities are identified by the project team members who will be performing the work as identified in the work breakdown structure or requirements. The schedule must also include the dependencies between activities, and there are several different ways that activities can relate:

- *Finish to start (FS) dependency.* The second activity cannot start until the first activity has completed. For example, a person cannot start a car until after the key is put into the ignition.
- *Finish to finish (FF) dependency.* The second activity cannot finish before the first activity. For example, when baking a cake the oven needs to warm up (activity 1) before the cake can be made (activity 2).
- *Start to start (SS) dependency.* Both activities have to start at the same time. Using the same baking example as above, this could be starting the oven and preparing the baking ingredients at the same time.
- *Start to finish (SF) dependency.* The first activity cannot finish until the related activity has started. An example could be a security guard who cannot leave his shift (finish activity 1) until the next guard comes to work (start activity 2).
- *Lag time.* This is the minimum amount of time that should pass between the finish of one activity and the start of another. For example, you cannot frame a picture that you just painted until after the paint dries (the wait time is the lag time).
- *Lead time.* This is when an activity can start early such as starting to test a component before its build is finished.

A significant consideration when creating the initial project schedule is the concept of a critical path. A critical path is the longest sequence of activities in the schedule that must be completed on time for the project to complete on the due date. An activity on the critical path cannot be started until its predecessor activity is complete; if it is delayed for one day, the entire project will be delayed for one day unless the activity following the delayed activity is completed a day earlier.

Many schedules also include the resources required to perform the activities. Based on the work breakdown structure dictionary (which describes the components and activities of the WBS), activity list, and activity durations, a project manager can work with the team to estimate the resources required. This planning activity helps to determine staffing and material needs, to assess team member capacity to perform the work, and to identify clear ownership of activities.

4.3.2 Manage the Schedule

After the schedule is created it needs to be managed closely as it serves many purposes for a project manager, several of which are outlined below. The schedule organizes the work into logical groups so there is clarity around what work has to be performed, what the key relationships between the activities are, and when the work has to be performed.

- The schedule is used during the execution of the project to track progress and completion of activities against the milestone dates.
- More complex schedule tracking techniques such as earned value management allow for a project manager to gauge progress toward goals before they are due, which can identify trends and allow time for corrective action.
- The schedule can also be used to model the implications of changes, issues, or risks. If dependencies are set up properly a project manager can see the impact of moving one milestone on other related milestones and the overall schedule. For example, an issue with a deliverable causing a delay in one project could have a downstream impact on another deliverable on a related project in the program.

So far, this section has focused on the explanation of a project schedule, but a program schedule is created and managed using the same approach. A program schedule should be an aggregation and roll-up of key milestones and deliverables from the project schedules. The biggest difference is that the program schedule includes dependencies between projects within the program and projects external to the program. The sections below discuss how to apply the consultative approach, guiding principles to schedules whether they are project schedules or program schedules.

4.3.3 Diligence

Because a schedule is the primary tool in managing projects and the program, it requires a significant amount of diligence and maintenance. Having a high amount of rigor in managing the schedule and the activities yields many benefits for project managers and the program manager, several of which are listed below:

- An understanding of the work required to meet commitments includes any key dependencies or milestones that need to be met.
- Keeping the schedule milestones and dates updated provides transparency into expectations and gives team members something to work toward. Without due dates and expectations setting, the natural tendency of people is not to feel a sense of urgency and, therefore, milestones will be missed.
- The ability to manage the many moving parts of the project or program by coordinating activities and managing the expectations of team resources.
- The ability to measure progress against goals with early indicators of tracking so that corrective action can be taken for any activities that are trending negatively.

There are several ways that a project manager or program manager can utilize diligence in managing the schedule:

- *Creating a standard structure.* Utilize a standard taxonomy of work to allow for better organization of the activities within a schedule. For example, as Figure 4.4 shows, a schedule could be organized by the project phases, deliverables, milestones, and activities. Then a program schedule is a roll-up of the projects with their phases, key deliverables, and dependencies.
- *Planning the work.* Meticulous planning is important because the path to meeting program objectives gets defined early during the delivery life cycle. The program manager needs to ensure that all work gets accounted for and that every last dependency gets identified. I always say that "If it is not on my list then it does not exist," which means if we do not have activities listed on the schedule then we cannot track them to completion. Another good practice in creating the schedule is to identify the commitment dates and work backward from those dates to ensure there is enough duration for the

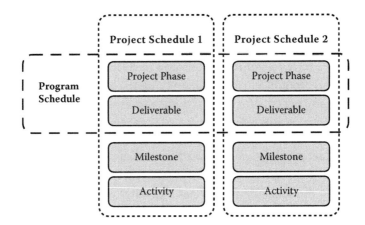

FIGURE 4.4
Schedule taxonomy.

work to be successful including any key relationships. Oftentimes commitment dates are set before the schedule is prepared; diligence in planning will avoid these situations where the program is set up to fail from the start (or it at least enables a fact-based decision on the implications). Lastly, schedule contingency may be considered for high-risk projects or activities that are not well defined.

- *Managing the work.* This means staying on top of all activities to understand what is needed to complete the work successfully. This can include ensuring that resources are available when needed, looking at upcoming activities to confirm they will start on time, and tracking the progress of work to meet planned dates. There are many methods that can be used to track progress such as asking team members for a percentage of work completion, tracking granular activities, or using techniques such as earned value management.
- *Understanding effects.* Having an updated and accurate schedule allows for the assessment of changes on a program. When a change is identified or a specific project has an issue or risk, the program manager can model the impact of the change on specific deliverables and benefits and can see the downstream impacts on dependent deliverables as well as on the overall program schedule.
- *Considering schedule contingency.* Every program has unexpected events and assumptions change, therefore a program manager should consider adding some contingency to the schedule. This is an effective way of providing some schedule "buffer" should problems arise (which often occurs).

- *Utilizing a tool.* There are many project planning tools in the marketplace that facilitate activity planning, resource alignment, and dependency management. Utilizing a tool can be beneficial, especially for larger and more complex projects/programs that have many activities, milestones, and interdependencies.

4.3.4 Attention to Detail

Not only do project managers and program managers need to be diligent with the management of their schedules, they also have to manage the details thoroughly. A poorly organized schedule will result in oversights of work, reactive management, missed dates, and a lack of clarity for program team members. Here are some specific areas to consider:

- *Identify accurate activities.* The activities in the schedule need to be accurate and include all deliverables required to meet the project goals. This can usually be confirmed by facilitating the activities from the team members who are doing the work and capturing all of the identified activities as well as their relationships to each other. Identifying and understanding dependencies are also important because this will help to understand the implications of changes, risks, and issues (which occur on all projects and programs).
- *Take all appropriate steps.* Because of the complexity of organizations and work, the project managers must make sure that they have all of the appropriate activities on their schedules. This could include processes or deliverables from a companywide delivery methodology or a set of steps that a dependent organization has to follow. Missing some of these steps could cause rework or unplanned work, so it is important to understand them early and include them in the schedule.
- *Monitor dates.* The schedule should be reviewed on a regular basis with special attention to dates of activities. This can include ensuring that activities will meet upcoming dates, following up on activities that are past due, and making sure that upcoming activities will begin on time.
- *Actively monitor the work.* Project managers and program managers also need to pay attention to how the work is progressing toward planned dates. An activity that has been at 80% for five weeks in a row may be an area of concern, but if a project manager is not monitoring the work closely, she may not recognize that trend.

It is imperative to have a detailed schedule and to manage its specific details. The project and program schedules serve to organize the work and mobilize the team to meet program goals effectively. If there is one place that a project manager needs to pay attention to the details, it is with the detailed schedule.

4.3.5 Transparency

With diligence and attention to the details of a schedule, the project manager can have an organized and accurate plan. The next step is providing transparency into the work that is being managed in the schedule. Transparency of work is critical to the success of the projects and the overall program for several reasons:

- The team understands what work they need to perform and when it is due. Without a schedule that is understood by the team, there is no clarity of expectations and commitments so dates will be missed or quality will be poor.
- Stakeholder expectations are managed as to when deliverables will be complete and what the logical steps of work are.
- Insights can be made early on how activities and deliverables are trending so any activities that are tracking off-target can be corrected in time.
- Key dependencies are understood so that any risks, issues, or changes can be understood as to how the projects and program would be affected.

There are two primary activities associated with providing transparency of schedule: communicating the schedule and trending insights. Communicating the schedule means sharing the key information with stakeholders. Table 4.6 outlines some of the different stakeholders, and the information they may be interested in as well as possible methods of providing them with insight.

Table 4.6 identifies that one of the most effective ways to provide schedule transparency is to generate a regular status report that includes milestones and progress against them. The status report is important because most project schedules are hundreds of line items, and a status report can be a rolled-up summary of key milestones. The status report can also show a visual display of deliverables (called a Gantt chart), which is a simple

TABLE 4.6

Schedule Stakeholder Communications

Stakeholder	Insight	Method
Project Team Members	• Understanding activities and milestones • Understanding what activities they are assigned to complete • Understanding dependencies of work	• Sharing the detailed schedule with the team. • Reviewing current and upcoming activities in project meetings. • Weekly status report.
Project Management	• Detailed and accurate schedule • Tracking of progress toward goals	• Weekly team meetings to understand activities and progress toward goals.
Program Management	• Key deliverables and milestones • Dependencies between projects	• Identification of key milestones and dependencies on project schedule; also adding them to program schedule. • Weekly status report.
Project Stakeholders	• Understanding of when deliverables are completed	• Project stakeholder meetings. • Weekly status report.
Dependent Projects	• Understanding of the key dependencies of work and progress against those milestones	• Linkage of those milestones in both project schedules and in the master program schedule. • Regular checkpoint meetings. • Weekly status report.

representation of the schedule. Then project, program, and stakeholder meetings can use the status report as a tool to review the activities and milestones. Section 4.12.3 goes into detail regarding the ways to use the status report to provide overall project and program transparency.

The second method for providing transparency is to get insight on schedule trends. Gaining these insights means managing the work at a granular level and tracking progress toward goals. Oftentimes on projects a milestone is recognized that will be missed one week before it is due, which is not enough time for the team to be able to take action to correct it. In order to avoid that scenario the project manager and program manager need to have early indicators of activities that may be heading off course. One of the best techniques for obtaining this insight is called *earned value management.*

Earned value management requires work be broken down into small entities so that each granular item can be "earned," and therefore the project manager can gauge progress based on the aggregation of completed granular tasks. For example, if a deliverable has 10 tasks associated with it and 4 are completed (and we assume they are equally weighted in effort), then that deliverable has earned 40%. If the plan has the deliverable expected to be 50% complete by now, then the deliverable is behind schedule, and the team can discuss how to bring it back on track.

Although earned value management does take some time in planning, it is effective at helping the project manager and program manager to understand their progress against goals. For example, program managers can know the trending status of a milestone when it is only 15% or 20% complete. There is significant documentation and also several tools in the marketplace that can help with earned value management. Two good examples include the *PMI Practice Standard for Earned Value Management* (2011) and *Earned Value Project Management* by Quentin Fleming and Joel Koppelman (2000).

The use of earned value techniques can oftentimes be a culture change in organizations because they require an additional level of rigor for the team that includes more granular planning and tracking of activities, detailed monitoring, and detailed time tracking. Program managers should consider the maturity of the team and organization before introducing these techniques. For example, in an organization that only tracks time at the project level, the program may want to consider tracking activity completion against time instead of based on actual hours reported in the time system. The best way to introduce these techniques would be to pilot them on a small project to demonstrate the value and then propose the concepts to senior management to get their buy-in.

There are other advanced techniques of providing schedule transparency including critical path analysis and program evaluation and review technique (PERT). Complex program teams should evaluate the techniques against the needs of stakeholders and maturity of the program to determine what appropriate methods to use for schedule transparency and analysis.

4.3.6 Single Sources of Truth

The schedule is one of the most important single sources of truth for any project or program. There are several project and program aspects for which the schedule is the authoritative source including the following:

- *Deliverables.* The schedule should have all of the deliverables iden-tified in the WBS and the associated deliverables and activities required to complete the work.
- *Milestones.* The commitment dates for the deliverables as well as dates for significant points or events in the project. These are what get rolled up from the project schedule into the program schedule.
- *Dependencies and sequencing.* The relationship of deliverables is crit-ical to have in the schedule as it provides clarity as to how the work is performed, in what order, and what would happen if any changes are introduced. These dependencies and sequencing may also drive the critical path of the project and program.
- *Progress against milestones.* One technique is for the schedule to track progress against dates as a percentage of completion.
- *Activity ownership.* The primary accountable resource associated with each deliverable should be identified in the schedule as well as the WBS dictionary and RAM. Note that most scheduling packages enable resource identification.

Because the schedule is the authoritative source of critical project and program information it is important to keep it maintained and as accu-rate as possible. It is also important to ensure that the team is consistently viewing this information to help manage expectations around milestone dates, ownership, and progress against goals.

4.3.7 Fact-Based Decisions

As noted in Section 4.3.4 the schedule is the primary source of many facts, which can be used in different scenarios to facilitate decisions or provide insight. Several of these scenarios where facts are used are identified below:

- *Scope of work.* Because the schedule is based on the WBS and con-tains all key activities and deliverables, it can be a good source for further refining the scope of work. These facts can be helpful in deci-sions regarding the scope of work and proposed changes.
- *Duration of work.* The schedule also identifies activities and their associated dependencies which then result in a total duration for work. These facts around timelines are needed in decisions and anal-ysis of duration. For example, in organizations that have a standard

release calendar there may need to be a decision about which release to plan for project work to be included. Based on the overall duration of the schedule including dependent activities, the project team can determine the right timing of the final release for the work.

- *Assessment of impacts.* During the life cycle of a program there are always risks, issues, and changes. It is imperative that a project manager or program manager be able to model the impact of these things as they get identified. Because the schedule includes all of the work with its duration and dependencies, a fact-based analysis can be performed to understand the implications of a change on a particular deliverable with the overall schedule. This analysis can help to facilitate conversations around impacts, options, and decisions as to how to proceed to take action against the risk, issue, or change.

These scenarios demonstrate how the guiding principles are interrelated. Having attention to detail and diligence around the authoritative project information allows for the facts to be accurate and then transparency provides the insight into these facts, which can then be used to facilitate effective decisions.

4.3.8 The Ships

The ships are also relevant to managing the schedule, especially given that the schedule is the authoritative source of important project and program information.

- *Ownership.* The schedule identifies the deliverables and the associated owner for each one. Documenting these forces the program team to determine ownership of work and then clearly state the named owner for each deliverable. Also, by identifying all project work in a detailed schedule a project manager is demonstrating accountability for all the deliverables and activities required to meet commitments.
- *Stewardship.* Because stewardship means to have responsible planning and management of resources, the schedule can therefore be considered the overall steward of scope and work. A project manager can demonstrate stewardship of work by having an accurate and updated schedule that is shared with stakeholders.
- *Leadership.* The project schedule is a great tool to support leadership because it allows a project manager to provide direction and

structure to the work to allow the team to progress toward a common goal. Demonstrating leadership means having a clear vision of what is required to meet the goal as well as the path to achieve these goals; these are the tenants of an effective project schedule.

4.3.9 Simplicity

Inasmuch as a program schedule has to contain all program work organized into projects, phases, deliverables, activities, and tasks it is common to see a schedule that has hundreds or even thousands of line items. Having a large schedule does not mean the concept of simplicity does not apply. In the case of a program or project schedule, the guiding principles of simplicity could mean organizing the plan so that it is easy to find and understand the work to be done. Utilizing a taxonomy like the one depicted in Figure 4.4 is a good way to organize the work. Other techniques for trying to simplify a schedule can include the following activities:

- Organizing work into logical groups that could be specific types of deliverables, project phases, or particular technology releases
- Indenting activities that are lower in the taxonomy to see the distinction visually between groups of work
- Color coding specific types or groupings of deliverables to have them stand out when the schedule is viewed
- Utilizing filters and grouping features in the schedule to allow stakeholders to filter different characteristics such as upcoming activities, a specific resource name, or a specific grouping of work

The key to having an easy-to-understand schedule is to organize the work in such a way that it is easy to understand the layout and find specific work. Poorly formatted or disorganized schedules make it difficult to find anything or understand the work (and thus there is a conflict with the guiding principles of attention to detail and transparency).

4.3.10 Taking a Customer Approach

As with the other program deliverables, the schedule has many different customers of information that need to be considered. Some examples of these customers are identified below:

- *Program team members.* Team members need to understand what activities they are assigned to and the expectations for when the activities need to be completed. They also need to understand dependencies of the activities so as to have a better context for the work they are performing.
- *Dependent projects.* In a complex program or portfolio of projects the dependent projects are also customers and need to understand the dates and progress against their dependent activities as well as any changes that affect those dates.
- *Management.* The progress toward milestones and commitments is important information that is shared with management, sponsors, and other stakeholders.

By recognizing and understanding the different customers of schedule information, a project manager or program manager can ensure that the information is accurate and meets the needs of his customers.

4.4 FINANCIAL MANAGEMENT

The last section focused on the schedule of project and program work with all known activities, and therefore the next logical progression is to focus on financial management, which is the cost of the resources required to perform these activities. Projects and programs typically have four different types of costs which are outlined in Table 4.7.

Each of the cost types listed above needs to be identified, managed, and tracked at a detailed level over the duration of the program. There are several processes required to manage program financial information, which are illustrated in Figure 4.5 in the context of a timeline for a program comprised of three projects. Project 1 is well under way and has completed estimates, incurred some costs to date, and has a forecast for the remaining work. Project 2 is almost finished and only has a small amount of forecast cost and time remaining. Project 3 has not yet started but does have an initial estimate. This example is very common for a program to have projects at different points in the life cycle at any given time.

1. *Initial estimates.* This process was covered during the work intake section (Section 4.2) in that the initial views of program costs are

TABLE 4.7

Types of Program Costs

Type of Cost	Description	Examples
Resources Internal	Resources from within the organization sponsoring the program	• Employees such as project managers, business analysts, or testers
Resources External	Resources from outside the organization sponsoring the program, including specialized contractors	• Fixed bid and variable contracted resources • Consulting companies • Software integrators
Material Costs	Purchases of materials required to complete the program	• Hardware and software for technology projects • Materials required for construction projects
Operating Costs	Additional expenses incurred to run the program	• Travel for resources • Office supplies • Training fees

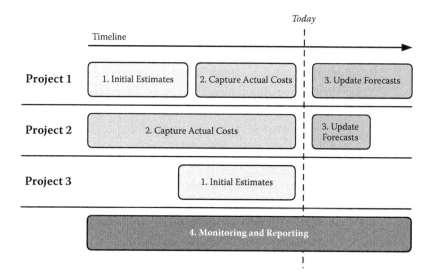

FIGURE 4.5

Financial management process.

identified during the estimation processes. There are many different methods of estimation as well as an iterative process of evolving forecasts and accuracy.

2. *Capture actual costs.* This is the process of accounting for spending that has already occurred on the program including resource time, vendor invoices paid, and purchases made.

3. *Update forecasts.* This process is focused on looking at remaining costs and ensuring that the assumptions made during the initial estimate are still true and that the remaining forecast is accurate as well. Otherwise the remaining forecasts need to be updated with new cost numbers.

4. *Monitoring and reporting.* This is the process of constantly monitoring the financials of a program and reporting on summaries and trends of the information. There are several types of tracking that can be used:

   • *Cost reporting.* Because programs are made up of many projects and are usually comprised of resources from many divisions, there are often many ways that the financial information needs to be organized and reported. For example, the program manager may want to see financials grouped by month and project, whereas an organizational lead may want to see cost broken out by organization and resource type. A finance resource may want to see information broken out by cost type. The cost data should be captured in such a way as to provide flexibility in reporting and organizing the data.

   • *Variance reporting.* It is important to be able to track variances from original estimates because usually the projects are funded based on those estimates and therefore expectations are set to what it will cost to deliver. Variance reporting can compare actual costs to planned costs for a particular month or the overall forecast against the original budget.

   • *Estimate at complete (EAC).* The EAC of a project or program is the total forecast spend of a project, and takes into account prior and future costs. It is calculated as the sum of the actual costs spent to date (ATD) plus the forecast remaining, also known as "estimate to complete" (ETC). The ATD, ETC, and EAC can then be compared to projections of expected trends of financial spend to determine if the forecast is tracking as expected or if it is trending in a different direction. Note that the technique listed above is one of several ways to calculate EAC and this works best for programs using earned value techniques.

Because programs have limited funds for completing their work as well as many costs associated with them, the guiding principles are very relevant and applicable to successfully managing financials.

4.4.1 Diligence

Diligence is essential to accurate financial management because of the need to manage the many factors involved in program costs including numerous resources with different rates and program allocations, multiple purchases, and tracking of invoices. There are many reasons for having diligence in managing program financials:

- Providing the team with a clear understanding of cost drivers and the assumptions behind the forecast costs
- Tracking of spend to date to understand how much of the original budget has been spent and therefore how much money is remaining to complete the work
- Managing the financial information well to identify changes and update forecasts accordingly
- Measuring progress against budgets with early indicators of tracking so that corrective action can be taken for any forecasts that are trending negatively

Diligence is required across all stages of the financial management process as shown in Figure 4.5. Not staying on top of any one of these areas could result in oversights, and the overall cost required to complete the program will not meet expectations.

- *Initial estimates.* As described in the work intake section, the initial assessment and identification of work are important to have accurate estimates inclusive of all known work. Diligence may also mean putting in some management reserve (known risks) or contingency (unknown risks) to hold money for unplanned costs. This should be considered especially on large programs, which have a high probability of financial volatility.
- *Capturing actual costs.* Identifying and accounting for all costs as they are incurred. This could mean utilizing time-tracking systems for resource costs, collecting invoices as they get paid, and understanding additional purchases such as hardware or software.
- *Updating forecasts.* Regularly revisiting assumptions and drivers of costs to confirm that the remaining costs on the program are accurate. This ties into having a robust change management process to identify and estimate the effects of changes.

- *Monitoring and reporting.* Constant monitoring of financial costs and forecasts to understand and report on any variances. These insights will allow project managers and team members to take corrective action, which can bring the forecasts back in line with expectations.
- *Maintaining a finance calendar.* Most companies have a standard cadence of when financial information is needed and reported. A program team should have a calendar with these activities plus the program-specific financial activities so that the entire team is aware of key dates in the process.

4.4.2 Attention to Detail

Once a program team is diligent about identifying and tracking financial information, it is then important to get into the details of the finances. It is in the details where the financial insight and trends are found and can be used to manage the work properly. There are several areas where the details need attention:

- *Checking your math.* First and foremost, it is imperative that the financial tracking be set up in a way to ensure that all costs are included and that the math works. Finances are managed in spreadsheets many times, and it is common to see math errors or mistakes in formulas (such as missing a row or column of information). These errors could result in reporting incorrect financial information, which could be misleading or even have a negative impact on the program results.
- *Understanding the details.* The financial information should be broken down into granular line items and also organized into different groupings for context. Then a program manager needs to dig into the details to understand what the costs are and how they align to the program and project work. A project manager or program manager should be able to explain where every penny of the project or program is forecast against and how they are tracking to those goals.
- *Understanding variances.* After the financial details are understood a program manager then needs to track and understand any variances that arise over time. Examples can include a purchase that was planned but got delayed or additional unplanned resources charging a project. In any case, time should be spent to understand variances

because action will need to be taken. In the case where the purchase got delayed and there was a favorable variance, the future forecast will need to increase to account for the purchase. In the case where additional resources charged the project, it may be that work will get done earlier, and the future forecast will go down.

It is common for large programs to set up a program office that has a financial management function. This function should provide the details and analysis to support the program team in understanding the details and stakeholders with key financial insight.

4.4.3 Transparency

Financial transparency is extremely important to be able to manage the resources and costs required to deliver the program. There are many cost drivers and forecasts are constantly being updated thus transparency provides a way to get a real-time understanding of what the forecast looks like so corrective action can be taken if necessary.

The introduction to Section 4.4 identified the concept of estimate at complete as the sum of actual costs to date and remaining forecast, also known as estimate to complete. Figure 4.6 shows two examples of how these concepts can be used to provide transparency into project and program forecasts.

In Example 1, we can see that the program has been incurring actual costs much lower than planned and has a positive variance of "A" at the

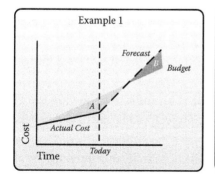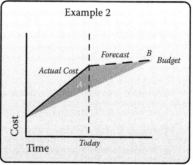

FIGURE 4.6
Estimate at complete examples.

current time. However, if we look at the remaining forecast, we can see that it is expected to increase at a much faster pace than the planned budget and will ultimately result in an unfavorable variance of "B" when completed. This is a classic example of a program that underspends at first because all of the costs get pushed to later months. Without this transparency of future costs, stakeholders may incorrectly think that the program is tracking to be under budget and spend that amount (A) on other things.

In Example 2, we have the opposite scenario where the program has been spending more money than it had planned and has a negative variance "A" at the present time. However, because that program has the insight and transparency of this financial information, the program management team has taken corrective actions to slow down the pace of spending so that they will come in on budget when the program is completed (point B).

Both examples highlight cases where the point-in-time understanding of variances is different than the final outcome of the program. Without the right level of transparency, the incorrect actions could have been taken, and the wrong expectations would have been set.

The concepts described here are basic tracking techniques and can be evolved by using additional earned value management techniques that also align schedule and financial tracking because of the natural relationship between the work managed by the schedule and the associated costs. Some of the key indicators that track schedule and cost are listed below:

- Cost performance index (CPI) is an indicator showing the efficiency of the utilization of the resources on the project and is calculated by earned value (EV)/actual cost (AC).
- To complete cost performance index (TCPI) is an indicator showing the efficiency at which the resources on the project should be utilized for the remainder of the project and is calculated by (Total Budget – EV)/(Total Budget – AC).
- Schedule performance index (SPI) is an indicator showing the efficiency of the time utilized on the project and is calculated by (EV)/Planned Value (PV).
- Schedule performance index (TSPI) is an indicator showing the efficiency at which the remaining time on the project should be utilized and is calculated by (Total Budget – EV)/(Total Budget – PV).

4.4.4 Single Sources of Truth

Because there are many types of costs that need to be managed, a program team needs to identify the single sources of truth for the financial information. There may be different systems or sources for different cost types but a program team will need to identify these sources and aggregate them to have one single source of program-level information. Examples of the different sources of program financial information may include the following:

- Original estimation templates with documented assumptions and drivers of cost.
- Tracking for vendor invoices and payments.
- Contracting system that has future payment schedules for contractors tied to delivery milestones.
- Purchases from a procurement system.
- Resource time tracking. Different organizations within the same company often have different resource time-tracking systems.
- Rate sheets for resources and role types.
- Software licenses.
- Different funding methods and goals.

Each of these providers of information needs to be validated as the recognized source of truth for an organization inasmuch as there needs to be credibility with the financial reporting of the program. Once the sources of truth are identified (and validated) for specific financial information, a master system needs to be created to aggregate all of this information in one place. This central location will then provide the total financial information across all projects and financial aspects including the following areas:

- Original budget
- Estimates and assumptions
- Actual costs spent
- Forecast costs remaining

This central source of truth should also be flexible in its tracking to provide multiple levels of reporting and organizing the information. For example, a program manager may want to see financial information by project, or a division lead may want to see all program costs associated with her area.

4.4.5 Fact-Based Decisions

At the core of the financial management function is using facts and insight to make decisions regarding the direction of the project and programs. All company executives have decisions to make regarding how to invest capital and what priorities will get funding each year. Understanding what each program will cost is a major contributing factor to how companies choose to spend their investments in technology and programs. Therefore, once a decision is made around funding, project and program teams need to do everything they can to meet the commitments of the work with the budget allocated to them.

As explained in the transparency section, programs need to understand both what their costs incurred to date have been as well as their remaining forecast. Having this insight plus diligent tracking of financials allows program managers to utilize facts to make key financial decisions, such as in the following examples:

- *Spending changes.* Facts concerning what comprise the costs incurred can allow a program manager to understand places where the program may be spending too much money such as vendors and software. These facts can then be used to make decisions about the continuation of these activities or to find less expensive alternatives.
- *Funding decisions.* Facts about a program having a forecast higher than budget can allow planning regarding ways to reduce future spending or discussions on obtaining additional funding. Obviously early insight and diligent tracking of financials should allow a program manager to understand these trends early enough to take action and avoid the difficult situations where she needs to ask for more funding.
- *Impacts of changes.* As issues, risks, or changes arise that can affect a program's forecast it is important to understand what the specific financial impact will be so that decisions can be made on how to fund the impacts. Programs also need to understand the impacts of changes to their business case, as too much additional cost could make the original cost benefit analysis (CBA) no longer attractive.

Programs make commitments to their sponsors on a budgeted spend, therefore, they must be diligent and transparent in the management of financials and utilize facts to make decisions that allow them to meet these goals.

4.4.6 The Ships

The ships are applicable to financial management because of the importance that finance has on program resources and commitments. Program managers need to have a heightened awareness and possibly even a different perspective on the management of financials by recognizing that they have ownership, stewardship, and leadership roles.

- *Ownership.* A program manager needs to feel ownership for every penny of the program regardless of where and how it is allocated. This means treating the funding as if it were his own money and making sure that it is being spent in the most effective ways on the most valuable resources and activities. Having this mind-set may avoid some of the waste and inefficiencies that programs have and may also result in making different resource decisions as well.
- *Stewardship.* Program managers need to recognize that they are stewards of the company's money and resources. If a program gets funding that means the company believes the program is a priority for the use of its money and is entrusting the program manager to manage it properly.
- *Leadership.* In the context of financial management, leadership means understanding how the program money is being spent, and when trends are identified, the program manager provides direction about what action to take to correct it. This may mean having to facilitate decisions regarding the prioritization of scope, resources, or schedule of the program to better align with financial goals and benefit targets.

4.4.7 Simplicity

Program financials can be very complicated and comprised of many granular parts such as hundreds of resources, consultants, vendors, and purchases. However, it is important to consider the guiding principle of simplicity and look for ways to make the digestion of financial information easy on stakeholders. There are several ways that financial information can be simplified:

- *Group costs into similar categories.* Organizing financial information into categories may help to understand the information better. These categories can be the projects within the program, resource types, organizations, or cost types. By using summarized categories, data can be compared month over month so that trends can be identified. Then the stakeholders can get into more detail on specific concerns versus trying to sort through all of the information in one view.
- *Use charts.* Charts are an effective way to display financial information and trends such as comparing actual spend to forecast spend over time. These visual representations can be helpful in stakeholders seeing the trends instead of trying to interpret them from spreadsheets of information.
- *Compare data.* An extensive amount of financial information is usually overwhelming to people, and it becomes hard to understand what the information is trying to tell us. One way to make sense of it all is to compare the information to something. For example, actual cost does not mean anything unless it is compared to budgeted cost. Program teams should look for baseline forecasts as a comparison point for future forecasts.
- *Use formatting.* On spreadsheets or presentations, formatting can be used to highlight important information, separate groupings of numbers, show totals, and also to structure the information. Often times the focus is on the numbers and not the presentation of the numbers, and this technique should not be underestimated as to the value and simplicity that it can provide.

4.4.8 Taking a Customer Approach

The customers of the program are expecting that the program meets its commitments including the costs that were agreed to up front. So by using the other guiding principles (e.g., diligence, attention to detail, transparency, and fact-based decisions) a program manager is taking a customer-focused approach to manage the financials effectively and meet customer expectations regarding cost.

Beyond the overall customers of the program, there are also many other internal customers of program financial information that need to be recognized and satisfied:

- *Program team members.* Team members are probably the ones who came up with the initial estimates (bottom up) so they need to understand how they are tracking against their initial forecasts and what assumptions have changed. Keeping the team members apprised of the financials also allows the program manager to hold them accountable for their initial estimates and assumptions.
- *Department managers.* Department managers need to understand their costs, forecasts, and resource allocations against the projects and the program as well as any variances against budgets.
- *Vendor management office/procurement.* Many companies have organizations that manage and report on contracts and purchases so these organizations will need to understand costs and invoices.
- *Finance division.* Most companies also have finance divisions that manage the program financials and also capitalize projects and depreciate assets. These divisions need to understand how project and program work relates to specific costs so they can correctly account for the work.
- *Managers.* Managers and sponsors are concerned with the overall spend and forecast as well as how the program is tracking against approved budgets on an annual basis and a multiyear basis for the CBA.

4.5 RESOURCE AND CAPACITY MANAGEMENT

Managing program resources is very similar in approach to managing the program schedule and financials. In fact they are very much related as Figure 4.7 highlights. The schedule is the collection of activities that resources work on with their associated duration, and the cost is comprised of resource rates applied over the duration of the work. So although this book separates the functions for purposes of explaining them individually, it is important to consider them as interrelated and manage them as such.

There are several functions that need to be considered when managing resources including resource planning, capacity management, and resource management, which are shown in Figure 4.8. Resource planning includes the identification and management of resources with the right skills required to perform the program activities. Capacity management is the alignment of the supply of these resources with the demand for those

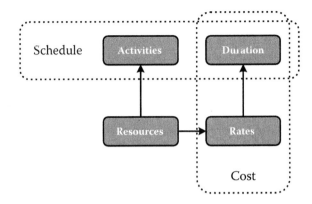

FIGURE 4.7
Resource management context.

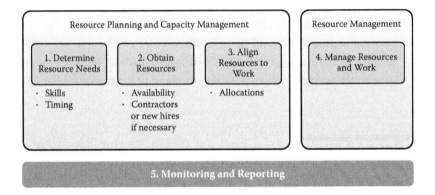

FIGURE 4.8
Resource management functions.

skills across a program. Lastly, resource management then involves the management of the resources as they perform the program work. These functions of resource management and capacity management are very much related and are discussed together in the remainder of this chapter. Note that the term "program resources" refers to more than just people and also includes materials required to complete the program. Because the majority of program resources are human resources and there is a lot of complexity in managing labor, this chapter focuses mainly in that area.

There are a few key activities related to the planning and management of program resources that are outlined below:

1. *Determine resource needs.* Resource planning occurs early in the business case planning and during program initiation and is when the program team determines what resources are required to perform the planning, support, and execution activities including the specific skills of the resources and the duration of time for which the resources are required. This is also known as the resource demand and is tracked in a resource management plan.

2. *Obtain resources.* Programs will work with the internal organizations that own the resources to identify available resources with the right skills needed to perform the work. This is known as the *supply of resource.* Sometimes new employee hires or external contractors are needed to fill resource gaps on programs. This can be when specialized skills are needed and do not exist or in the case where a certain type of resource is not available within an organization

3. *Align resources to work.* Once the resources are identified and aligned to the need, the resources need to be allocated to the work properly in the human resource management systems. These systems are used to match resource supply and demand and also involve making sure that program resources are fully allocated to work. This is important as it becomes the tool by which program teams can optimize the use of resources. Additionally, there are complex resource planning functions including resource leveling where resources are aligned to granular tasks with effort, which is then analyzed to determine how many hours of work each resource is assigned per week. The work is then "leveled" to ensure that the work is evenly distributed over time. Note that leveling resources will have an impact on the schedule which needs to be considered during planning.

4. *Manage resources.* Once resources are assigned to the program they need to perform the work properly to meet program commitments. The management of resources can include ensuring that team members have all the tools necessary to perform their activities against the expected plans. Having a program resource plan is an effective technique for managing resources.

5. *Monitoring and reporting.* Resource information can be monitored and reported across all functions. These reports can include an identification of the program resource needs, a roster of human resources with their allocations to specific program activities, an inventory of open roles with their progress toward being filled, and an inventory of materials.

Resource management is a critical function because the resources on a program are the keys to its success. Programs can be considered nothing more than activities performed by resources to meet a common goal, so without the proper attention to the resources a program will not be successful. Keshishian and Walkow (2010, p. 6) say it well when they say that "The greatest technology does not make a project successful. People do." Therefore, each of the guiding principles should be utilized when planning and managing program resources.

4.5.1 Diligence

Having diligence with regard to resource management means that resources with the right skills are properly aligned to the work and can complete the work when it is expected. There are many reasons why this is important and why diligence is necessary:

- Managing resources well means understanding what resource skills are needed and when resources are needed so they can be aligned properly to meet program objectives and commitments. A resource management plan is a good tool for identifying and quantifying all of the labor and materials required to complete the program.
- Because of the growing complexity of program work the skills required to perform each activity have evolved and have become very specialized. For example, there may be the need to have resources who are experts in business rule technology, data modeling, or integration technologies. These resources are usually hard to find and in high demand so diligence is necessary to identify these needs and make sure these resources are obtained. These resources are also expensive, and program managers need to recognize and plan for these higher-cost resources.
- Without understanding and staying on top of resource needs, there is a risk that resources may not be available on time, which will have an impact on project and program schedules. Early identification of resource gaps allows time to obtain resources, which is especially important because many companies have long processes for hiring new employees or procuring contractors.
- Diligence should be considered around contingency planning and succession planning because some resources are critical to the

program or have a hard-to-find skillset. In either case losing those resources will have an impact on the program and so proper consideration should be given to those scenarios.

Because resources are at the core of delivering programs, a diligent approach to planning and managing them is directly aligned with the success of the program. Several techniques for applying diligence to resource management are listed below:

- Proper planning is needed up front to align resources to activities and understand the specific skills required. It is not enough to say that a program needs 20 technology developers inasmuch as there are dozens of different types of programming languages and one developer is not the same as another.
- Once these resources are identified, the program team needs to work actively with the resource managers to line up the appropriate resources to be available when needed by the plan. Because many organizations have resources assigned to multiple projects at the same time, this requires a good amount of coordination by the resource managers as well as lead time to finish up prior assigned work. Resource managers are accountable for making sure that their team members are fully utilized and allocated at 100% across programs or projects and proper planning is needed to make sure they are not over-allocated or under-allocated to work.
- If resources with the necessary skills are not available, the program team and resource managers will have to start the process for hiring new employees or lining up external contractors. These processes usually require long lead times and many steps (e.g., forms, approvals, etc.), therefore they should be started as early as possible.
- After resources are engaged and assigned they need to be managed well to ensure that they have what they need to meet their program commitments successfully. Diligence has to continue through to the end of the work and not stop once resources have joined the program team. This includes contingency and succession planning in the cases where there are key program resources working on critical activities. Possible techniques could include having some team members shadow the work of the key resource or proper documentation of program work.

4.5.2 Attention to Detail

There are many aspects to program resource management that require a focus on quality and details. These include resource skills, allocations, and timing:

- *Resource skills.* It is important to understand the details of the types of resource skills that are needed for each activity. Having the best fit of resource skills to assignment needs is a significant contributor to the success of the activity and therefore also to the success of the program.
- *Resource allocation.* A program team needs to understand all of the resources allocated to the work. Having resources with too low an allocation could mean slowdown of work and impacts on the schedule. Resources that are over-allocated can also become overwhelmed and produce low quality of work or miss dates. Also having resources who are not fully allocated across the program means that they are not being managed efficiently, and costs are being spent on unproductive work.
- *Start dates.* A program team needs to pay attention to upcoming start dates of resources to ensure that resources are available and ready to start on time. This also means completing any logistical planning such as acquisition of a computer, allocation of working space, and access rights to systems or tools. Any delays in starting resources on time can have an impact on scheduled work.
- *End dates.* Just as important as looking at starting dates of team resources is to look at ending dates. Because program financials are based on assumptions of resource allocations, any extension of resources beyond planned end dates will affect the financial forecast of the program.
- *Understanding risks.* Another detail to be aware of is the risk of losing specialized resources who are working on key activities. This can include paying attention to the morale of these key resources. Turnover is common, especially on programs that span over multiple years. It is a best practice to recognize who these "single points of failure" are and determine contingency plans should those resources decide to leave or get pulled onto higher-priority work. An example can include an exit interview to understand reasons for leaving with a checklist of key questions to ask. Another technique could be signing a non-disclosure agreement (NDA) to protect intellectual property if a resource leaves.

Because of the need for attention to resource management, many programs will have dedicated team members who focus on resource planning and capacity management. This attention is important because of the amount of coordination and planning required to manage the multitude of resources working on programs.

4.5.3 Transparency

Programs need to have transparency regarding their resources because resources drive all of the work to meet program commitments. Therefore any challenges with resources have a direct impact on the program and need to be understood as early as possible so that appropriate action can be taken.

There are several ways that a program can provide transparency into the resources needs and trends:

- *Program organization chart.* Having a visual display of the organization is helpful to manage expectations of how the program team roles relate to each other, especially in a matrixed environment (which is how most programs are organized).
- *Roles and responsibilities.* Documenting the specific roles on a program (such as a business analyst) and their associated responsibilities (such as eliciting requirements) is useful to show the team members what each role is accountable for on the program.
- *Resource roster.* Program teams should maintain a roster of all resources working on the team. This should include information such as the project being allocated to, organizational alignment, and start/stop dates. Open roles should also be tracked in this roster. By having a roster of all resources (and keeping it updated), a program can have transparency into what resources are working on the program and where there are any outstanding roles that need to be filled. Programs may even want to consider tracking the skills of resources in this roster as well.
- *Supply and demand.* It is important to have tracking regarding the need for resources with specific skills to work on program activities (demand), and the resources who are assigned to those activities (supply). One way of tracking this is by having resources assigned to activities within the project schedule. It is important to keep this

alignment updated in the schedule and not view this as a one-time planning activity.

- *Time reporting.* Earned value management is a useful technique to track and report resource time against control accounts which then roll up to the broader program. This time is then tracked against expected effort to gauge progress against plans.
- *Resource leveling.* A program can build on the alignment of resources to activities by doing resource leveling, which is an advanced resource management technique that looks to balance the conflicting interests of program schedules with the availability of resources. For example, a program deliverable has the need for three full-time resources working for four weeks but there is only availability of two resources with the needed skills. This is important insight to have because a program manager can then determine if he can get the third resource or if he has to slow down the work and expand the schedule to meet the availability of resources. Most project management tools have these advanced features to align resources to work and provide transparency into resource loading and leveling.

4.5.4 Single Sources of Truth

Program teams should have a master inventory of all resource information and keep that information updated often. This information can be stored in project management systems or even a spreadsheet, but the key is to keep it current given that programs have many resources who are always starting and stopping and the information can get outdated very quickly (i.e., diligence principle). Examples of the types of program resource information that should be tracked are listed below:

- *Resource names.* The name of the individual or a "TBD" (to be determined) if the role is open but not yet filled by someone.
- *Resource role.* There are many different roles on programs, and this element is used to show which resources are aligned to which role type. For example, roles can include business analysts, project managers, developers, or testers.
- *Resource type.* Resources should be classified as either internal employees or external contractors.

- *Organizations.* This element shows the organizational alignment of internal resources.
- *Project assignments and allocations.* The allocation of each resource to the specific project assignments provides insight into how effectively resources are allocated to the work and the specific work assigned.
- *Start and end dates.* Documenting when resources are expected to start and stop project assignments. These dates are important for resource managers to understand inasmuch as they are responsible for ensuring continuity of work for resources across several different projects, which have different and often conflicting milestones.
- *Rates.* Some resource systems include "role rates," which are average hourly rates for each resource type such as a developer or project manager. This becomes useful for estimating the cost of projects when there are effort estimates, which can then be multiplied by the role rates to determine a forecast cost.

Oftentimes, companies will have project management information systems or human resource management systems, which should automate and track many of the resource elements listed above. These are valuable tools on programs, which have hundreds of resources because of the complexity required to manage all the parts of a program. They also allow companies to perform resource management functions across projects and nonproject work such as allocation of resources and calculations of cost.

Additional authoritative sources of truth for resources on a program can include having a program organizational chart and role descriptions. Both of these are needed to explain the structure to stakeholders and manage expectations for the team members.

4.5.5 Fact-Based Decisions

Having the program resource information tracked and kept current as suggested in the previous section will enable many fact-based decisions that the program manager will need to make during the course of the program. These decisions will have implications for program costs and schedule and therefore having the right facts will enable the optimal decisions to be made for the program:

- *Alignment of resources to work.* Understanding to what work resources are assigned helps to facilitate decisions regarding the ability to meet commitments. In cases where resources are not allocated sufficiently to work, decisions can be made on acquiring more resources or changing schedules.
- *Financial implications.* Cost is a major consideration for any program or organization, and sometimes decisions need to be made to reduce the cost of programs to meet financial pressures. Information regarding the rates of resources and the allocations of those resources to specific activities or programs may be needed to make decisions around less expensive resources or the re-prioritization of less costly work.
- *Impacts of changes.* As changes arise on programs, managing resources effectively means decisions can be made concerning how to staff the new changes and the impacts on other work already assigned to the resources.
- *Contingency planning and succession planning.* Given the large number of resources on programs and the volatility of the business environment, there are often changes in program scope and resources. Therefore program teams need to plan for the loss of key resources as well as proactively create succession plans. Having detailed information about resource allocations and skills will allow program managers to work with resource managers to plan accordingly and identify contingency plans.

4.5.6 The Ships

Managing resources also requires program managers to utilize the guiding principle of the ships. These are significant concepts because people are at the core of successful program delivery, and the program manager has a primary role in the management and motivation of the people on the team.

- *Ownership.* Program managers need to take ownership over the program work and commitments and the only way to do this is to recognize that the resources are what make a program successful. Therefore, they need to do everything in their power to make sure that the resources can be successful at meeting their goals, which could include facilitating decisions, removing roadblocks, or acquiring additional tools. Ownership of the work also means providing clear roles and responsibilities to the program resources so there are clear expectations and alignment of the work to peoples' roles on the team.

- *Stewardship.* Program management is not just about meeting the work commitments. Program managers can also be stewards of people's careers by helping them to grow and learn new skills. This can include pairing up team members with more experienced resources, providing additional training, or putting people into stretch assignments and providing the right amount of support. Also, resources do their best work when there is an optimal fit of the work to the resources' skills and interests. Therefore by helping people to get a better alignment of work to their skills, the results are improved team morale, better quality of work, and higher probability of meeting program goals.

- *Leadership.* Program managers play a very important leadership role on the team. They need to motivate and champion resources to meet objectives and then recognize them when they are successful. They also need to provide direction, facilitate quick decisions, and resolve issues as they get identified. Program managers should view their organization as an inverse pyramid where they work for the team members and enable the team to do their jobs effectively which, in turn, will make the program successful. Leadership also means being able to influence senior management and other program stakeholders through proper escalation, communication, and facilitation of decisions.

4.5.7 Simplicity

As with other functions, program teams should use the concept of simplicity regarding resource management. Because there are so many resources involved on programs with different allocations having simple views of resource information will make it easier for stakeholders to understand the resource landscape of the program. There are a few ways that program teams can apply simplicity to their resource management function:

- *Resource reporting.* The resource reporting should be easy to view and understand, therefore organizing techniques such as filters, sorting, and grouping are good techniques to use. These allow stakeholders different views of the information such as resource managers being able to view their resources or project managers being able to view the resources working on their projects.

- *Capacity views.* One important aspect of program management is aligning resources (supply) to the work (demand) appropriately, and therefore capacity views need to be simple and easy to understand. This can include clear depictions of what resources are needed and when they are needed.
- *Resource management.* By using a human resource management system, program teams can leverage existing functionality and screens to make the management of resources easier. These systems usually come with the ability to assign resources to work and run different types of reports such as resource allocations, time-tracking, or resource skill inventories.

4.5.8 Taking a Customer Approach

There are many customers that need to be recognized and considered in the resource management function. The program manager should be aware of these customer relationships and do everything she can to maximize the value to those customers.

- *Resources.* Resources are the ones performing the work, therefore they can also be the customers of information or dependent activities. Program managers and project managers need to make sure that the team members have everything they need to be successful, which can include proper tooling, training, software licenses, or help from other team members.
- *Resource managers.* Managers need information regarding the status, allocation, and performance of their resources. They then use this information to manage resource supply and conduct performance management activities. Because their goal is 100% utilization of their resources they need the proper estimates and allocations to plan accordingly across the many projects and programs they support.
- *The company.* The company sponsoring the program can also be a customer of the skills of the program resources. If a program can increase the skills and effectiveness of the resources the company will benefit from productivity in future programs. This perspective is consistent with the stewardship principle of looking to grow the skills of the team members.

4.6 VENDOR MANAGEMENT

With regard to technology programs, a vendor is an organization that offers contingent labor or products. Vendors play an important role on programs and using them has many benefits. Some of the reasons that programs use vendors are outlined below:

- *Niche skills.* Vendors may have expertise in specific skills that some companies lack. This may include new technologies such as mobile computing or recent trends in development techniques such as agile programming.
- *Experience.* As technologies become more complex and companies' strategies become more aggressive, the size of programs increases in both cost and schedule. Not many companies have people with experience in managing extremely large programs, and therefore vendors may be brought in to help deliver the work.
- *Risk sharing.* Because large programs are complex and risky, partnering with a vendor can help share in the execution risk and therefore provide a higher probability of program success.
- *Additional resource capacity.* Vendors can play an important role in managing resource capacity by providing additional resources on demand as programs need them. This approach reduces the time it takes to recruit, hire, and bring on new resources. This approach also saves the overhead cost and time of hiring resources as well as reduces the risk of downsizing resources when programs finish and there may not be additional work. Lastly, it avoids costs for companies such as benefits, bonuses, and training.
- *Cost advantage.* Some vendors offer an offshore model where there are resources who work for lower wages on certain program roles. The most common example of this is technology programmers, who can cost around 50% less than local resources. As programs become more expensive to deliver, companies are constantly looking for ways to keep development cost down and offshore vendors provide an opportunity to do this. Using vendors for certain skillsets also allows a company to focus its efforts on building strategic skillsets such as business analysis and project management.
- *Product offering.* Some vendors also offer technology products or tools, which can increase the delivery of solutions on programs.

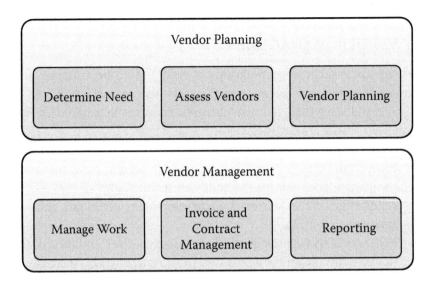

FIGURE 4.9
Vendor management functions.

There are several functions under the category of vendor management. These functions are highlighted in Figure 4.9 and can be organized into activities performed for vendor planning and activities performed for vendor management:

- *Planning: Determine need.* To start, program teams should identify the need for vendors. Program teams can initially consider a "build versus buy" analysis to determine if it makes sense to bring in a vendor (buy) or perform the activity with internal resources (build). There are several reasons that a program team would identify a need including additional resource capacity, a reduced unit cost of resources, specialized technology skills, or a product offering that aligns with the program solution.
- *Planning: Assess vendors.* Once vendor needs are identified, program teams need to scan the marketplace for vendors, publish a request for a proposal (RFP), and then assess the results against a set of criteria. Most organizations have a vendor management division, which can help with the process of identifying and comparing vendors. These divisions many times will also have a qualified vendor plan where vendors have been prescreened; this will expedite the process of searching for vendors.

- *Planning: Vendor planning.* Once a vendor is selected, the program then needs to complete the planning required to bring the vendors onto the team. This can include completing vendor contracts, planning workplace logistics, and obtaining access for vendor resources to building security or internal systems.
- *Managing: Manage work.* After planning is completed and the vendors have joined the team, program teams then need to manage the vendor work. This can include monitoring of the planned activities, performing quality reviews of the work, and signing off on specific deliverables. Many organizations use a contract management plan that outlines the method for which a specific contract will be administered and executed. This document can be used as a framework for managing the vendor work.
- *Managing: Invoice and contract management.* Another aspect of managing the vendor relationship involves financial and contract management activities. This can include tracking forecast costs, confirming contractual milestones are completed, and processing invoices (either fully or with some retainage based on partially completed work). Another aspect to contract management is closing the contract properly to ensure that all contract requirements have been satisfied.
- *Managing: Reporting.* Inasmuch as programs are increasing their use of vendors, a reporting function becomes significant to provide insight into vendor productivity, performance, and quality. This can include reporting on overall costs, resource utilization of offshore and onshore resources, and contract information. These reports can be used for vendor performance reviews and discussions as well as determining if the vendor should remain on the qualified vendor list of the company.

Because of the many benefits of using vendors they are becoming more integrated with programs and therefore need the same approach with regard to the guiding principles as any of the other program management functions.

4.6.1 Diligence

Using vendors on programs means spending "external" company money on contracts and resources and so it is important to make sure that the money is being spent effectively on getting the most value. Therefore program managers and team members should be diligent across each function of vendor planning as well as during vendor management.

- *Determine need.* Programs need to take the time to appropriately identify when they need vendors. Vendors are expensive and should not be used for every activity, but on the other hand they provide an additional level of execution experience and confidence for a program. Program teams should document a set of criteria for when to use vendors and then apply that to their program to determine the risk areas that could potentially see improvements from using an outside vendor. Then a buy-versus-build analysis should be performed using the set of criteria and ranking each of the options to see which is optimal for the program.
- *Assess vendors.* Because vendor engagements can last a long time and cost a significant amount of money, it is important to spend the time up front analyzing vendors' capabilities and conducting due diligence. The analysis of vendors should include several factors such as prior experience, references, and cost. Once a vendor is selected, the contract should also be written in such a way as to align payments with milestones and performance (also known as service level agreements) or a cost-plus contract that allows for additional payments. This type of contract gives incentives to the vendor to meet the interim milestones and provides a way to track progress. There are four types of cost-plus contracts:
  - *Cost plus fixed fee:* pays a predetermined fee.
  - *Cost plus incentive fee:* pays fees based on meeting or exceeding performance targets.
  - *Cost plus award fee:* pays a fee based on vendor's work performance (which may be subjective).
  - *Cost plus percentage of cost:* pays a fee as vendor's costs rise.
- *Vendor planning.* Diligence should also be taken to prepare to bring on the vendor because time will be lost if the vendor team joins the program team and does not have the right access or tools needed to perform the work. This can include acquiring computers, system access, or access to key team members. Many companies have procurement management plans that describe the procurement processes from developing procurement documentation through contract closure. These plans should be used during vendor planning.

- *Manage work.* Proper management of the vendors is important to understand how the work is progressing against the contracted milestones or service level agreements. Techniques for monitoring can include signoffs at specific points in the process, reviews of work, or aligning the vendors with team members who can report on the progress and quality of work. It is important to create a partnering relationship with the vendors for mutual benefit and not just treat the vendor as a subcontractor. A partnership where both teams feel responsible and work well together will have an optimal effect on the program results and quality.
- *Invoice and contract management.* Contracts have very defined terms, milestones, and payments so program teams need to stay on top of managing the contracts and invoices properly as sometimes there are even fees or penalties written into the contracts for not meeting the specific terms. Plus it is good business to pay vendors on time as work gets completed per the contract.
- *Reporting.* Most companies that use vendors also have procurement or vendor management divisions, which require reporting on progress against contracts as well as vendor demographic information such as roles, rates, and location of vendors. Programs need to be accurate and thorough in their management and reporting of this information as it may be used for strategic decisions within the company.

4.6.2 Attention to Detail

Building on the approach of diligence, program teams also need to manage the details and quality with respect to vendors. Programs can spend a significant amount of money with vendors and entrust them with critical components of delivery, therefore program teams need to pay attention to all aspects of the vendor relationship to ensure successful completion of work. Several areas to pay attention to are explained below:

- *Quality of proposal.* Pay attention to the details and quality of the proposal returned back regarding the program work as it may be a good representation of the quality of work that the vendor will perform. For example, if the proposal from a vendor is disorganized and messy, the work performed may be of the same quality.
- *Quality of work.* Although vendors should be coming in with relevant experience, they are not always aware of the specific nuances of the

company culture so it is important to perform reviews of their work to ensure that it aligns with company standards, processes, and guidelines.

- *Contract specifics.* There are details within the contract that need to be understood. For example, some contracts document the initial cost but then apply a tax or percentage of cost for expenses, and program managers need to account for these costs in their forecasts. Other specifics to understand can include the start and end dates of resources, rates for overhead (such as relationship managers), and statements about intellectual property.
- *Contractual milestones.* Most vendor contracts are set up so that payments are tied to specific milestones. Therefore these need to be tracked and understood to determine if the milestones are being met and when the invoices should be paid.
- *Vendor end dates.* End dates should be well understood as well inasmuch as vendors who extend past dates will cost the program money that may not have been planned. Contracts that are ending may also require the program to bring on additional internal resources or to prepare for additional work.

Vendor management requires diligence and attention to detail, and therefore, most large programs have a vendor management function within their program office. This function performs that tracking and reporting of vendor information and performance and works with the related organizations within the company.

4.6.3 Transparency

Programs need to track and manage vendor work and provide the same level of transparency as they do when managing internal resources. There are several ways of providing insight concerning vendor engagement on programs.

- *Performance of vendors.* Because vendors are contractually obligated to perform work, there needs to be a level of transparency into their progress. The most common way of doing this is to set up the contract to pay based on interim milestones and performance and then have team members perform reviews of that work to confirm that it meets quality and contractual expectations. These milestones should be tracked and reported in the same way as the schedule-tracking techniques mentioned in Section 4.3, "Schedule Management."

- *Vendor information.* Programs need to understand the use and effectiveness of vendors. Also, as companies are trending to use vendors more on programs and other parts of the business, they are becoming more interested in understanding how and where vendors are used within their organizations. For both reasons, there needs to be a high level of transparency regarding vendor information, which can include contract specifics, assignment of work, and financial information.
- *Financial management.* Along with the tracking of work performed by vendors, there also needs to be tracking of the financial components associated with the contract including invoices, spend-to-date, and overall expected spend.

4.6.4 Single Sources of Truth

As mentioned in the previous section, it is important to have transparency into vendor information, which includes having a master inventory of this information. This inventory becomes the single source of truth for all vendor information on the program and should include the following details:

- Contract information includes vendor name, contract name, vendor resource names, key contact information, description of work, milestones, payment information, and end dates for work and type of contract (fixed/cost, plus/time, and materials).
- Financial information includes any actual costs incurred, the forecast amount, projects that are paying for the vendor and budget or expense codes that need to account for the spend.
- Invoice tracking includes the vendor name, invoice number, amount charged, and progress against paying the invoice.
- Assignment of work for vendor resources. Similar to managing internal resources, vendor resources need to be aligned to specific projects and deliverables on the plan.
- Reporting and aggregation of vendor information, which can be grouped by vendor name, project, spend to date, forecast spend, budget code, and period of time such as month or year.

As with the other master inventories, the vendor information needs to stay current and updated as information changes or as new vendors get contracted. Keeping the information current and applying diligence will result in a high level of transparency and insight for the program and the company.

4.6.5 Fact-Based Decisions

Deciding to use a vendor and then which one is the best fit for the program requires making difficult decisions that need to weigh options against requirements. Section 3.6.1 discussed a method called quality function deployment (QFD) whereby options can be evaluated. This is a good technique for assessing vendors because it compares options against a set of weighted criteria. This process will allow programs to determine the criteria by which they want to assess vendors as well as the priority of those criteria. There are several criteria that should be considered when evaluating vendors:

- *Alignment of proposal to solution.* How well does the response from the vendor match the requirements in the RFP?
- *Experience.* The years of experience that a vendor has in completing similar work. This should also include the experience of the team members who are being proposed by the vendor.
- *Cost.* Overall cost of the vendor proposal including resource costs, expenses, and other purchases.
- *Schedule.* The schedule that each vendor is committing to in its proposal.
- *Company profile and risk.* This can include years in business, number of clients, and financial stability.

By using a set of weighted decision criteria, program teams can make the comparison of options for a fact-based decision. It may also make sense to include an internal proposal of what it would take to complete the work with existing resources, as a point of comparison or for a build-versus-buy analysis. Other fact-based decisions regarding the vendor management function can include decisions regarding vendor performance against contracted milestones and financial analysis of work performed for cost spent.

4.6.6 The Ships

Utilizing vendors on programs expands the program team and the number of stakeholders who are engaged in the success of the program. They also require additional management attention, and therefore the ships become important to use in the vendor management approach to the program.

- *Ownership.* A program manager owns all of the work associated with the program, which means also treating the vendor resources as part of the team. This includes providing clear roles and responsibilities, inviting vendor resources to team meetings, and providing the vendors with everything that they require to successfully perform their work.
- *Stewardship.* Program managers can also look for ways to leverage the experience of the vendor team to grow internal team members. This can mean partnering team members with experienced vendors or even writing into the contract milestones regarding knowledge transfer. Stewardship can also mean considering that the company's money is being spent on vendors and looking for ways to maximize the effectiveness of the relationship and thus get the best "bang for the buck."
- *Leadership.* Program managers need to demonstrate leadership to vendors as well as internal resources. This can include providing direction and seeking to motivate them. Some vendors with offshore resources come from cultures where customer praise is held in high regard so it should be considered as a tool to use. Another leadership trait is recognizing the vendors as part of the team and not to perpetuate an "us versus them" mentality. This is especially important in roles that may have been outsourced to lower-cost organizations, and there is a feeling of resentment within the company.

4.6.7 Simplicity

There are several areas regarding vendor management that can utilize the concept of simplicity. The first is in the case where a vendor is contracted to provide a solution. Programs need to understand the complexities of the vendor solution and make sure that it does not add more complexity to their environment, which will result in higher maintenance and support costs. For example, if a program solution requires one module to be updated, but the vendor creates several new modules, these will now require support and maintenance. The solutions need to be scalable to allow for changes, easy to integrate into existing solutions, and easy to maintain once the program is completed. Often, companies become enamored of new products and do not consider the total cost of ownership or ease of maintenance in a solution and then end up with a product that is costly, difficult to change, and requires a lot of support.

Within the vendor management function of a program office, the vendor reporting also needs to be simple to understand and read. Inasmuch as there are many different stakeholders involved with the management of vendors, the reporting should be clear and easy to understand. For example, the program team will need to understand costs by vendor for each project, the finance organization may want to see contract or invoice information, and the vendor management organization may want to see performance and quality information. The reporting capabilities should be flexible and organized in such a way as to make it easy for these stakeholders to obtain the information that they require.

4.6.8 Taking a Customer Approach

There are several different customer relationships involved with the vendor management function that need to be recognized and considered:

- *Reporting recipients.* There are different customers of vendor reporting including the program office, finance organization, and vendor management organization. Each of these needs to receive certain vendor and program information to perform their function effectively.
- *Senior management.* Vendors are becoming heavily aligned with company strategies, and therefore senior management will want to understand if they are getting the performance, productivity, and quality that they are expecting from the vendors.
- *Vendors.* Even the vendors can be considered customers on the program. They are customers of deliverables that are earlier in the delivery life cycle. For example, if a program is using programmers from a vendor, they are a customer of the requirement and design deliverables, which need to be documented in such a way that the vendor team can successfully complete its activities.
- *Vendor organizations.* In companies that have vendor management or procurement divisions, those divisions will want to understand the performance of the vendors against their goals. They are customers of this information from programs and will use it to maintain qualified vendor lists and contract information.

4.7 CHANGE MANAGEMENT

Change management has several meanings for programs and in the industry. For the purposes of this chapter, I am using the term to refer to managing the changes in program characteristics (e.g., scope, cost, schedule, etc.) and not preparing an organization for changes that the program will introduce. Change management is an important function of running a program because it puts controls around the scope, schedule, cost, and benefits of a program. Many times programs end up missing schedules, cutting scope, or coming in over budget, and the blame is put on the original planning or estimation. I believe that rigorously managing to original assumptions and estimates combined with a structured change control are the key to success of a program. Assumptions will change as the program progresses but it is critical to assess the impacts of those changes and then make informed decisions on how to proceed. This process helps to manage expectations of stakeholders instead of seeing what appear to be random fluctuations in program schedule or cost.

Figure 4.10 shows the four main steps in the change management process. This process begins once scope, schedule, cost, and benefits have been baselined and the program has started to perform work against its commitments:

1. *Identify change.* The first step in the process is to identify that there is change that has an impact on a project. Changes can be identified in many ways but generally speaking, having diligence and attention to detail across all functions of program management will reveal

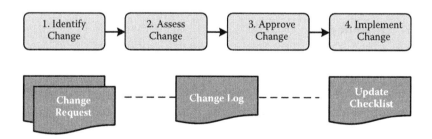

FIGURE 4.10
Change management process.

changes early. A change request is then submitted that contains the key information regarding the change such as the background on the change, requestor information, and project(s) affected.

2. *Assess change.* The change is then assessed for full impacts across the program including scope, schedule, resources, quality, benefits, risk, and cost.

3. *Approve change.* Changes with impacts are then brought to an approval group to review the implications and make a decision on what to do with the change. This group is often referred to as a "change review board" (CRB) and is comprised of key program decision makers. Some change control processes have thresholds by which the changes need to be brought to the CRB. For example, if a change has no effect on schedule but costs $10,000 a program may choose to just process it without going to the review board. Changes with larger impacts may get reviewed by an executive group such as a steering committee. Decisions on whether the change is approved are then documented. All of the change information is stored within the change log.

4. *Implement change.* If a change is approved it then needs to get implemented, which includes updating any of the program's single sources of truth. For example, if the change has an impact on the schedule then the project schedule and other master schedules need to be updated to reflect the change.

Change management is essential to meeting program commitments and warrants tight adherence to the guiding principle because it requires rigor, provides transparency of work, enables fact-based decisions, and is a critical step in keeping the master inventories current.

4.7.1 Diligence

There are many changes that arise on programs and each change has implications across the various characteristics of a program including schedule, dependencies, cost, benefits, resource, and scope. Because changes can have significant implications on those program dimensions diligence is needed across all steps in the change management process.

- *Identify changes.* Having diligence in management of the program will result in understanding what the baseline is for scope, schedule, cost, and benefits as well as early identification of changes to those areas. Once changes are identified, program teams should have a change request form and an inventory to track changes (also known as a change log). Both of these documents should capture key information on the changes, impacts of changes, and disposition regarding what to do with the change.

- *Assess change.* This is an important step because all implications of the change need to be understood, otherwise there may be unplanned work later or a poor decision could be made based on incomplete information. Changes need to be thoroughly examined by team members who are familiar with the solutions and who can identify the nuances that will result from the change.

- *Approve change.* This step requires ensuring that key stakeholders are involved in the process and can make informed decisions as to the implications of accepting or declining the change. All decisions should be documented with key participant information, disposition, date decision was made, and reasoning for the decision.

- *Implement change.* Lastly, attention is needed once the decision is made to update all of the relevant program information. A checklist should be developed and then used to track that all sources of truth get updated accordingly. These should include the project and program schedules including dependencies, cost forecasts, risk log, resource plan, vendor inventory, and any associated scope documents.

4.7.2 Attention to Detail

Quality and a detail orientation are important in the change management process because of the need for accurate information to make informed decisions regarding changes and their impacts. There are several areas with regard to change management to which program managers should pay special attention:

- *Accuracy of submission.* An effective change management process starts with having an accurate change request submission. Having a change request form or template with specific questions will force submitters to think about all affected areas as they document the request. It may make sense to list every scope item or organization within a program and make the requestor consider if there is an impact or not instead of having a blank form where the requestor may forget about a certain area.

- *Assessment of impacts.* The identification of all program impacts of a change allows for better informed decisions. Therefore it is important to identify and document all of the details with regard to the change. Having diligence across the other program functions will make this easier because details around schedule dependencies, cost and benefit drivers, resources, and scope will all be accurate and up to date. The requestor and change team should compare the change to these program and project deliverables to model the effects of the proposed change.
- *Keeping all master inventories updated.* Single sources of truth are one of the key guiding principles and are critical to keeping program information updated and visible to the team. Therefore, the details of a change need to be understood and appropriately updated in these master inventories as decisions get made and the changes get implemented. Without the attention to detail on updating these changes, the inventories will very quickly get outdated and become useless.

4.7.3 Transparency

The change management process is fundamentally about transparency, including the identification of changes and the effects on the program. Programs need to provide visibility of the change request throughout the entire change management process to ensure that everyone involved with the program can easily find out the current status of any changes, the assessment of impacts, and the reasons for approving or rejecting specific changes.

There are several primary techniques that a program can utilize for providing insight into the changes.

- *Change log.* This is a master inventory of all changes that stores information about the change request, identifies the progress of the change, tracks the impacts, and documents the decisions and reasoning. This becomes the one-stop-shop for any stakeholder looking for information regarding the changes.
- *Impacts.* The impact assessment houses all of the impacts of a change for all dimensions of the program as well as the assumptions and reasoning for the estimations of cost and schedule. This allows a team member to see the details of the change effects and the thinking behind them.

- *Updated documents.* A checklist should be used to provide transparency into which other project or program deliverables need to be updated based on the decision about the change. This checklist then should be managed to provide transparency into which documents are updated and which are not yet updated.
- *Reporting.* Changes should be included in key program reporting such as status reports or executive briefings so there is clarity about changes including their effects and the decisions made regarding them.

4.7.4 Single Sources of Truth

To provide the transparency suggested in the last section, a centralized program change log should be created and maintained with all of the relevant change information. This becomes the single source of truth for all change information and an historical record of decisions and assumptions. Table 4.8 identifies the key elements that should be captured and managed in a change log.

Change management is also closely aligned to the principle of single sources of truth in that the process results in the updating of the other program sources of truth such as the schedule, financials, benefit plans, resource roster, and vendor inventory. Change management is critically important to keeping the program sources of truth "living" because the process identifies all possible changes to those master inventories and then tracks that they get updated properly.

TABLE 4.8

Elements of a Change Log

Element	Description
Change Request Information	Basic information about the change request such as requestor name, date submitted, description of the change, project affected, risks, and impact if change is not approved.
Impact Information	Listing of impacts by type (cost, schedule, resource, vendor, scope) and areas (business, testing, development, etc.). This also should include assumptions made regarding the impacts.
Change Request Progress	Tracking what the status of the change is along with the change management process.
Decision Information	Capture information on when decision was made, by whom, what the decision was, and what the reasoning was for the decision.

4.7.5 Fact-Based Decisions

The change management process is set up to enable fact-based decisions regarding whether to accept the change and what the ramifications of doing so would be. There are several important facts needed to make those decisions.

- Impacts on the scope of the program.
- Impact on schedule based on the activity affected and the subsequent dependencies within the project and the other related projects.
- Impact on the financial forecast including the cost of the change and the cost associated with any of the downstream impacts. For example, if the change extends the schedule and resources are required to remain on a project longer, those costs need to be identified and included.
- Impact on the benefits expected to result from the program.
- Impacts on resources including internal and vendors. These facts need to be understood because the resources may have been planned to go to other projects and those projects would be affected or would need to find different resources.
- Impacts on the risks of the program that may require additional mitigation or planning to avoid.
- Impacts on the quality of work such as defects, removed scope, or the end product to the customer.
- Facts concerning what would happen if the change were not accepted. In some cases the changes cannot be avoided and therefore it becomes less about making the decision and more about understanding the effects on the program.

Once all of the facts are gathered an informed decision can then be made regarding how to proceed with the change. Note that based on the effects of the changes, there may be different levels of management that need to make the decision. There are several options that can be made regarding how to proceed with a change.

- *Accept.* The change (and identified implications) can be accepted in full or partially. The program team would then have to implement it and update all appropriate program documents and inventories.

- *Accept with tradeoffs.* The change would be accepted but other tradeoffs would be required. For example, the change could cost an additional $100,000 and the decision would be to cut another project scope to recover the $100,000 which would result in no overall impact on the program.
- *Defer.* The change is still warranted but the decision could be to defer it to another time. This is usually the case when there are not available funds or tradeoffs and the change is not a high priority.
- *Reject.* The decision is not to proceed with the change. This could be in the case where the impacts of the change are greater than the impacts of not performing the change. Note that there may be impacts of not doing the change, so these impacts need to be considered as well.

In any case the decision should be documented along with the reasoning for the decision and who made the decision.

4.7.6 The Ships

Program managers need to use the ships to properly and diligently manage all of the changes for the program as they arise. Change management is a key control process and therefore requires all three of the ships to be effective.

- *Ownership.* Program managers need to manage the changes through the entire change management process including making sure that change is implemented properly. There are many steps requiring diligence and attention to detail and therefore having an ownership mind-set will enable these to be performed optimally as opposed to assuming that someone else will do it.
- *Stewardship.* Regarding change management, stewardship means identifying and understanding all impacts of the change request. By understanding the changes and impacts, the program team can effectively steward the company's money and resources by enabling better fact-based decisions as to how to proceed with the change and what the tradeoffs would be.
- *Leadership.* Leadership is also required during the change management process. This can include being a champion for the process, helping to facilitate the fact-based decisions, as well as escalating

and communicating impacts to senior management and other stakeholders especially in scenarios where tradeoffs have to be made. Providing this leadership will allow the change management process to be followed by the team members and provide the right level of senior management visibility.

4.7.7 Simplicity

There are many changes that arise during the course of any project or program and also several stakeholders involved in the process of assessing and approving them. Therefore, having a simple and easy to understand process is important to facilitate an effective change management process. Two areas that require an approach of simplicity regarding change management are the change request form and reporting.

- *Change request form.* This is the start of the process and the initial contact with the stakeholders so the request form needs to be simple to understand and easy to use. At the same time, the request form needs to gather enough information to allow for proper assessment of the change. There are a few ways in which the request forms can be simplified.
  - Adding instructions where appropriate (although the principle should be to create it in such a way that it is self-explanatory)
  - Having drop-down fields with prepopulated information
  - Not asking for similar information in several places
  - Grouping questions into logical categories and using formatting to section off these groups
- *Simple reporting.* The reporting of the changes should also be easy to use and understand for the different stakeholders. Therefore different reports or views should be considered. For example, the program management team may want to see the impacts across the program whereas the requestor may just want to understand what step in the process the request is in and what the decision was.

4.7.8 Taking a Customer Approach

There are several internal customers of the change management process that need to be recognized and understood:

- *Decision makers.* In order to make fact-based decisions on how to proceed with a change, the stakeholders making the decision will need information regarding the change and the quantified impacts. This information will need to be captured, organized, and presented in a way that allows these stakeholders to make optimal decisions. The change management team should understand how these stakeholders prefer information to be presented and then tailor the approach accordingly.
- *Requestors.* The people who submitted the requests for change are also customers of the decision regarding that change. They need to be communicated to effectively regarding the outcome of the decision and the reasoning behind it.
- *Affected areas.* The organizations and projects affected by the change are also customers of information regarding the change. They need to understand that there is a change that can affect them and be provided an opportunity to assess the change and identify effects in their area.
- *End customers.* Changes affect the end customers of the program, which can be organizational units or the end customer of the company. These people need to understand that there is a change and what that means to the program commitments that were made to them. Sometimes there may even be a debate with these stakeholders over whether the change is really a "change" or if it is a "missed requirement" which would depend on different perspectives. In either case it is important to acknowledge that the original estimates are being affected and it is the program's responsibility to identify and assess those impacts.
- *Sponsors.* If the change has a significant impact on program scope, financials, benefits, or schedule then the sponsors need to understand these implications. The program needs to present this information to them to outline the justifications and impacts.

4.8 RISK MANAGEMENT

Risk management is the identification, assessment, and prioritization of risks followed by coordinated and economical application of resources to minimize, monitor, and control the probability or impact of unfortunate events (Hubbard 2009). Inherent in any program is risk and as programs and their solutions get larger and more complex, so do the risks and their implications. Therefore risk management is a critical part of program management and does not mean creating a risk log early in the program and then storing it away for "later." Risk management has to be embedded within every aspect of program management because risks can occur anywhere such as in the following examples:

- Risk of schedule slipping due to an unplanned activity or resource contention that would have an impact on key milestones
- Risk that the solution does not satisfy the requirements
- Risk that the assumptions made in the original estimates were wrong and that the program is more costly than originally planned
- Risk that a vendor or product will not deliver a solution that aligns with the design and scope of the program
- Risk that resources are not available when needed by the program
- Risk of change in sponsorship for the program and therefore a change in scope or direction
- Risk that the company leadership or direction will change and that the program objectives will no longer align to the strategy (which is a very real risk for programs that span multiple years)

There are hundreds of books and materials on risk management techniques, but I simplify it down to the four steps shown in Figure 4.11 inasmuch as all of the techniques are optimally effective when used with the guiding principles.

1. *Identify risks.* The first step is identifying and documenting the risk. This requires performing regular assessments of the program and diligent monitoring of the master sources of information to identify possible risks.
2. *Assess risks.* Once risks are identified they must be assessed. There are many methodologies to do this that could weigh the risks, score them, or even quantify their financial impacts. There also needs to be analysis of the root cause to see what is driving the risk. Once the risks are assessed the program team needs to determine what to do with the risk and can take one of three actions: accept it, avoid it, or reduce it.

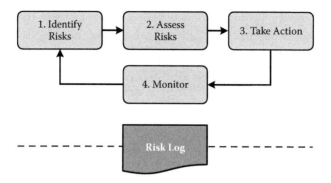

FIGURE 4.11
Risk management process.

3. *Take action.* Upon assessment and determination of approach, the program team then needs to act on the risk. Choosing to accept a risk may mean looking for ways to share it. For example, if a program is using new technology for the first time and the program team thinks there is risk that it won't work properly, they could decide to work with a vendor who may have experience in the technology to share in the risk.

4. *Monitor.* Risk management is a living process and therefore even if action is already being taken, the program team must continue to monitor it to ensure that it is having the desired impact on the risk. Also constant monitoring is important to identify additional risks as there are always new ones popping up.

4.8.1 Diligence

Program risks can be elusive and are not as tangible as a project schedule. The way to identify and manage risks to minimize their impact on the program is through constant diligence.

- *Identify risks.* Diligence enables the identification of program risks as early as possible which then allows for ample lead time to assess them and take action. The risks should be documented into a risk log or risk management plan. There are two ways to be attentive to risk identification.
  - Perform regular assessments of the program. This can include facilitating risks from team members on a regular basis or just stepping back to consider known information about the program and where the potential risks might be.

- Diligent monitoring of program sources of truth. By constantly monitoring program deliverables such as the program schedule, resource plan, vendor inventory, and financials a program manager can identify risks. For example, the schedule and financial models should provide insight to possible slippages of deliverables and costs.
- *Assess risks.* There are many methodologies to assess risks and their impacts. The techniques are not as important as the principle of being diligent to confirm that all impacts are identified as well as understanding the root cause of the risk. There are several methods that look to score or quantify the risks. Two examples are demonstrated below but note that just because something cannot be quantified does not mean it is not a risk that needs to be managed.
  - A composite risk index is calculated as the impact of the risk event (on a 1 to 5 scale with 5 being the highest impact) multiplied by the probability of occurrence (also on a 1 to 5 scale with 5 being the highest probability). By using the same criteria a program can normalize the risks by relative score and then determining low, medium, and high risk scores based on ranges from 1 to 25.
  - Another technique to quantify risk involves multiplying the percentage probability of occurrence with the financial impact of the risk. For example, if there is a 10% chance that a risk will cause a $100,000 program cost then the risk is calculated at $10,000.
- *Take action.* Once the risk is assessed and a root cause is understood, the program needs to determine the approach to take action on that risk. Whatever the action is determined to be, it needs to be managed with the same level of diligence as the deliverables on a project schedule. This means the actions need named owners with completion dates and that they get monitored on a regular basis.
- *Monitor.* The risk log can be the central repository of tracking the actions as well as storing all of the risk information. The key to risk management is the constant monitoring of the program to continue to identify and track risks.

4.8.2 Attention to Detail

Paying attention to the details will allow a program team to identify possible risks. There are several sources of truth for different aspects of the

TABLE 4.9

Examples of Understanding Details to Provide Risk Insights

Source of Truth	Risk Insights
Schedule	• Earned value tracking yields risks of not meeting interim milestones which could affect project milestones or program commitments • Dependencies and duration of activities force the critical path beyond expected commitment dates
Financials	• Financial tracking shows trending of actual costs coming in higher than budget or trending of future forecasts increasing
Vendor Inventory	• Vendor activities are not progressing against the contracted schedule milestones or scope
Resource Roster	• High resource turnover especially with critical team resources • Skills of resources do not match required skills for solution • Resource start or end dates are not consistent with demands of project schedule
Change Control Log	• A high number of changes can mean scope volatility and risk of meeting commitments

program and each can provide insight into possible risks as long as the program team is looking carefully. Table 4.9 provides some examples of sources of truth and possible risk insights that they can provide. In each of these examples, there are leading indicators of risk that should allow time to take action before they get realized, which would then turn them into issues.

The other area where detailed tracking is important is in the management of the risk log. Because the risk log is the master inventory of risks it needs to be managed with the same level of detail as the program schedule or financials. Therefore all actions need dates and owners and need to be followed up to ensure they get completed.

4.8.3 Transparency

Managing risks means having transparency into those things that can cause problems on the program, understanding what the effects of those problems can be, and being able to take action before those effects get realized. There are two primary techniques for providing insight into program risks:

- *Risk information.* A risk log should be used as a master inventory of all program risks; it includes information about the risk, the assessment of probability and impact, as well as any tracking of actions being taken. By having all risk information in one place it becomes the source of truth for all program risks and actions and team members will know where to go to find this information.

- *Impacts.* Transparency is also needed for understanding the impacts of risks so that appropriate action can be taken. This approach requires the diligence described previously to conduct the assessment of probability of occurrence as well as a detailed impact analysis on the different aspects of the program. Once this analysis is performed and assumptions are documented, it should be organized in such a way that the program team has access to it and can understand what the impacts are.

The guiding principle of transparency is also important for the risk management function because by having transparency in all aspects of program management, it will become easier to recognize where there are risks that require taking action. Information regarding risks that is not available or organized effectively may hide important clues regarding risks that could be missed.

4.8.4 Single Sources of Truth

Because risks can arise anywhere on the program and they require assessment and attention, program teams should maintain a central repository of risk information. This way program risks are captured in one location and can be reviewed and updated on a regular basis. Regarding the risks, the risk log should contain key informational elements that can be grouped into three categories:

1. *Risk information.* This includes key information regarding the risk, including the date it was identified, how the risk was identified, and a description of what the risk is and how it may become an issue.
2. *Assessment information.* This section of the risk log should include any quantitative analysis of the risks. The probability of risk occurrence should be documented along with relevant assumptions. Impacts also need to be documented and should indicate all affected areas and the impacts on cost, resources, schedule, and scope. There should also be documented assumptions regarding the impacts.
3. *Action tracking.* Each action in response to the risks needs to be identified with a named owner and a target date for completion. There should also be commentary on the status of the action.

By having a single source of truth for risks a program manager can provide transparency into the risks as well as facilitate effective analysis of the risks and decisions regarding how to act upon them.

4.8.5 Fact-Based Decisions

Identifying program risks is important and requires diligence as well as attention to detail to uncover them but once they are understood, actions are needed (otherwise it is no different than not identifying them). A program team needs to make fact-based decisions regarding what actions to take; however, not all risks should be considered equal and some require more action than others. Figure 4.12 organizes the types of actions based on probability and impact of the risk.

- *Low probability and low impact.* Inasmuch as these risks are both infrequent and with a low impact, the program should document them but spend more of their focus on other types of risks.
- *Low probability and high impact.* These are risks that may not happen but when they do will have a big impact. Because these have a low probability of occurrence the team should not spend too much time on trying to mitigate them but should have a documented contingency plan. This way, if the risk does get realized then the team will have a plan as to how to handle it.
- *High probability and low impact.* These are probably the most common types of risks where there are many possible risks but they do not have large impacts. These can tend to add up to larger risks so the

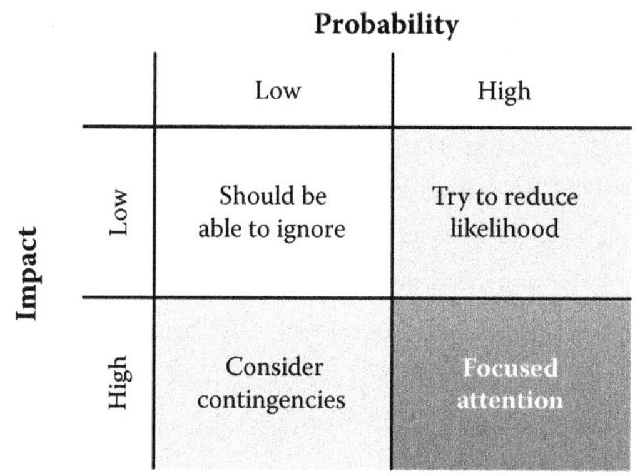

Probability

		Low	High
Impact	Low	Should be able to ignore	Try to reduce likelihood
	High	Consider contingencies	Focused attention

FIGURE 4.12
Risk actions.

actions should be to try to manage these as best as possible to reduce their likelihood of happening. The guiding principles of diligence, attention to detail, and transparency should help to minimize these types of risks.

- *High probability and high impact.* It is hoped that programs do not have many of these but these are risks that are very likely to happen and will have a larger impact on the program. Programs need to spend significant focus on these to look to minimize, avoid, or share these risks and may have to take significant measures to do so. In this case, facts are important so a team can articulate the impacts and present options to senior management to make decisions prior to the risk occurring.

Once the probability and impacts of a risk are understood, a program has the necessary facts to determine the appropriate course of action and make the right decisions. Because both probability and impacts can be hard to quantify and may be subjective, they may not be true "facts" but by using the team's best professional opinion they can be managed is if they were facts.

4.8.6 The Ships

Risk management is a difficult function to manage because it is a little nebulous and it is based on what could happen as opposed to more tangible functions such as cost, schedule, and resource management. Therefore it is important to have the mind-set from each of the ships to manage program risks.

- *Ownership.* By recognizing that program managers are accountable for the outcomes of the program, they will feel more ownership over the risks that could jeopardize those results. Then risk management becomes less of an administrative activity and more of a proactive tool to figure out what could go wrong and try to minimize or eliminate those things altogether.
- *Stewardship.* As mentioned before in this book, program managers are the stewards of the company's resources and are entrusted to deliver on commitments. Having a solid risk management structure will improve the probability of using those resources optimally on a program.

- *Leadership.* A good leader is proactive and understands where the risks are hiding so she can aggressively confront them. This may also mean facilitating difficult decisions based on events that have not yet happened which require influence and strong communication skills to articulate the impacts.

4.8.7 Simplicity

Program risks can be complex as well as have many different impacts across a program. Therefore simplicity is needed to ensure that the risks and their impacts are clear and well understood by stakeholders. Simplicity means that the risk log needs to be well organized and easy to understand, which can be challenging given that larger programs tend to have many risks that need to be documented. There are a few ways that a risk log can be organized to help simplify the information:

- Use categories and group risks by them. For example, stakeholders can see risks for a specific project or view all risks that have financial impacts.
- Sort and organize risks by severity so that stakeholders are not overwhelmed by all of the risks and can focus on the most important ones.
- Filter on risks that are not closed. It is good to have the historical documentation but once a risk has been avoided or mitigated then it can be moved out of view.
- Put the status of the risk action in the log with a date so team members can easily understand the progress against the risk.

Once the risks are organized into the risk log the other area where simplicity should be used is in the articulation of risks to stakeholders or senior management (especially if the team is proposing an action against the risk). It is not necessary to explain every detail of the risk or impact but the information should be presented in such a way that it is clear as to the risk and impact of not addressing the issue as well as the implications of taking action such as additional cost, resources, or schedule. Simplicity of the presentation will allow the stakeholders to focus on making the right decisions and not be confused about specific details or irrelevant information.

4.8.8 Taking a Customer Approach

As with every program management function, the risk management function also has internal customers that need to be recognized and managed. There are the obvious program team members who are customers of the risk information in order to manage their work. There are also the stakeholders who may be affected by the risks in one of two ways: either they will be affected if the risk occurs or they will be affected by the actions required to mitigate the risk. This could mean different work, additional resources, or changes in schedule. These stakeholders then become customers of the risk impacts and decisions that are made so it is important to involve them in the process and communicate the information to them so that they can plan accordingly.

Another set of customers is the stakeholders who will be making decisions on how to manage the risks. These people need to understand the information regarding the risk and how it will affect the program so they can make informed decisions. They then become customers of the impact information as well as the possible actions and implications of those actions. This information needs to be organized in a way that is meaningful to the decision makers and it needs to be accurate so that they can make the best possible decisions.

4.9 ISSUE MANAGEMENT

The last section was focused on program risks but when the impacts of risks are realized they then become issues. Program managers can plan with diligence resulting in the perfect estimate and schedule but if issues are not managed well, then the program team will have a limited ability to be successful because they will always be "fighting fires" and reacting to issues. Some examples of common program issues are noted below:

- A deliverable is taking longer to complete than was originally planned for in the project schedule.
- A dependent project is slipping its schedule which has an impact on the program's dependent milestones.
- A resource from another division is not available to start per the resource plan and work will be delayed.

- A key resource on whom the work is dependent is leaving the project.
- Unplanned costs are being recognized.
- The quality of the solution is not meeting expectations.

The issue management process is shown in Figure 4.13 and has similar steps as the risk management process.

1. *Identify issue.* When an issue is identified that usually means there is an impact on a program, which could include implications on scope, resources, financials, benefits, or schedule. An issue log should be used to capture the risk information.
2. *Assess issue.* As issues arise they need to be assessed for complete impact as well as options to resolve the issue. This assessment can involve multiple program team members who have insight into the impacts and who can offer solutions.
3. *Escalate issue.* Depending on the impact of the issue, an escalation path should be defined and used to alert management and facilitate any decisions needed to take action to resolve the issue.
4. *Address issue.* After action is determined, the issue needs to be aggressively managed and closed.

Having a solid approach to issue management is important given that issues always arise on programs. What makes some programs more successful than others is the ability to manage these issues effectively, which is where the guiding principles get used.

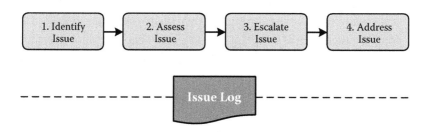

FIGURE 4.13
Issue management process.

4.9.1 Diligence

Once issues are identified they need to be resolved as quickly as possible to minimize the impacts on the program. One issue could derail months or years of work on a program and therefore being diligent with an issue management process is fundamental to managing programs successfully:

1. *Identify issue.* It is imperative that the program management team has current knowledge of the work and the operations of the program so that issues get identified early. Program managers and project managers should be conscious that identifying items trending toward becoming issues is a critical part of their jobs. Techniques such as earned value management, status meetings, management by walking around, and stand-up meetings are intended to facilitate information from the projects to identify risk areas and potential issues early. Issues need to be diligently captured and monitored in a central program issue log.
2. *Assess issue.* Assessing an issue requires diligent analysis to determine how the issue will influence the program as well as what the possible options are.
   - *Impact assessment.* A thorough impact assessment needs to be performed to understand the implications of the issue on program schedule, scope, resources, quality, cost, and benefits.
   - *Option analysis.* The team should also determine every possible option to resolve the issue as well as document the assumptions and implications of each option. The team should also determine a recommendation for which option to pursue and be able to justify it to management.
3. *Escalate issue.* A solid issue escalation path must be in place for the program which means that team members need to understand who they should raise something to and when they should raise it. A documented escalation path should clearly explain what thresholds require what level of management attention and to whom to escalate. Program teams should also strive to maintain a culture of honesty, quick identification of issues, and a focus on resolution (as opposed to blame). People who get punished for raising issues will not do that again, and in that environment, by the time the issues are known they have become much bigger.

4. *Address issue.* Once a decision is made on how to resolve the issue the actions need to be aggressively managed and closed. This also requires due diligence to make sure that the actions implemented do not cause additional issues and have the intended effect on the program.

4.9.2 Attention to Detail

Issues can have significant implications for a program and therefore paying attention to the details is required during the entire issue management process. Similar to monitoring for risks, the program team needs to focus on the specific details of the management tools to identify when an issue arises. For example, one missed deliverable dependency within a program schedule could result in the overall critical path moving out beyond the committed milestones, which would be a big issue on programs that do not have schedule flexibility such as a compliance initiative.

When an issue is discovered it is important to manage the specifics of that detail well in the master issue log. There are many details that need to be carefully documented and monitored including the following:

- *Impacts.* A thorough analysis of impacts is needed so that each area of the program gets considered for implications as to resources, costs, schedule, and scope.
- *Owners.* Every issue needs an owner so that the program manager knows who to follow up with regarding issue closure.
- *Dates of actions.* Each issue also needs a date assigned to it and this date needs to be tracked to confirm completion.

Without the attention to the details noted above the impact of the issue may be misunderstood or even become worse as time passes. It is the combination of diligence and attention to detail that will provide the earliest indication of a problem on the program and therefore the fastest resolution of that problem.

4.9.3 Transparency

The issue management function requires transparency regarding the issue information, the impacts of the issue, and the actions taken. Having a centralized issue log enables team members to capture the issue information along with the impacts and actions associated with those issues. For programs there should be an issue log for each project and then one at the program level. The project issue logs should contain issues specific to that project, many of which can be managed from within the project. The program issue log should capture those issues that span across projects or affect any of the program goals. The program manager and project managers need to keep the information in these issue logs updated so that any stakeholder looking at the issue log can have the most recent information.

Transparency within other program functions also allows issues to be identified earlier which in turn provides more time to take action on the issue. For example, good financial or schedule transparency will identify any areas of concern that may be going off course and will not have the expected result.

4.9.4 Single Sources of Truth

The issue log should be the single source of truth for all issues on a project or program and be monitored and updated regularly. There are many benefits of having a centralized issue log available to the team:

- Having a consistent method and place for the team to raise and document issues.
- Tracking and assigning responsibility to named team members for each issue.
- Analyzing and prioritizing issues easily.
- Documenting issue resolution for future reference and historical purposes. Those who do not pay attention to history are doomed to repeat it so it is always good to see what was done in the past with a similar issue.
- Monitoring the overall project health and status.

The issue log should contain relevant information regarding the issue. Table 4.10 outlines the key elements that the issue log should document and monitor.

By having a centralized issue log with the elements listed, there will be full awareness of all issues as well as what is being done to close them out. The key to good issue management is not just having the central repository, but also diligence in keeping the information updated and accurate.

TABLE 4.10

Elements of an Issue Log

Element	Description
Issue ID	A tracking number for the issue.
Issue Description	A description of what the issue was including how it was discovered.
Date Identified	The date that the issue was documented.
Identified by	The person who discovered the issue.
Issue Type	Categorizing the issue into types such as infrastructure, resource, financial, scope, or schedule.
Project	Which project or workstream is affected by the issue.
Impacts	A description of the impacts and their severity including scope, cost, financials, resources, and quality.
Priority	The urgency of the issue that prioritizes which issues need to be focused on first.
Assignment Owner	The person who is assigned with resolving the issue.
Resolution	The actions required to resolve the issue. Note that there may be several required and each should be tracked separately.
Target Resolution Date	The date to get the issue settled.
Comments and Status	Tracking of progress of the resolution toward the dates. It is helpful to put dates next to comments to see the full audit trail, for example, "October 1: Worked with resource manager to get additional resources with a target of October 6. October 6: Resources acquired and planned to start."

4.9.5 Fact-Based Decisions

Once all information has been captured on the issue and all the impacts have been assessed, the team needs to pull together options and a recommendation regarding what action to take to close out the issue. There are several facts that should be included in the option comparison and used to make a decision:

- Severity of the issue including the implication if the issue is not resolved and the timing of those implications. Teams should be able to answer what the outcome would be if the issue is not resolved in one week or one month.
- Specific impacts of the issue on the resources, scope, schedule, and quality of the project as well as the program.

- Implications of the options to resolve the issue as the actions taken can also have an influence on resources, costs, benefits, scope, schedule, and quality.
- Feasibility of the options as there may be constraints or challenges with the implementation of the action.

After all of these considerations are assessed with several alternatives and a recommendation is made, there needs to be a decision on how to proceed with the issue resolution. Because issues sometimes can get worse over time it is important to weigh gathering more information with making a rapid decision based on available information. Program managers need to present the information in a way that is meaningful to the stakeholders and so a decision can be made. That decision then needs to be documented and action commenced.

4.9.6 The Ships

Identifying and then resolving issues quickly is critical to successful program management. Issues that are identified late or that go unresolved can have significant implications on the ability for the program to meet its commitments. Program managers need to use the guiding principles of the ships to drive the appropriate actions regarding issue identification and management:

- *Ownership.* Program managers need to create a sense of ownership on the team regarding issue identification and resolution. Team members need to take accountability for meeting their commitments and therefore accountability of resolving any issues that have the ability to affect those commitments. Having accountability means not viewing themselves as just reporting the issues but rather driving them to closure.
- *Stewardship.* As a steward of the company's money, program managers need to recognize when issues can affect program commitments and do everything in their power to close them out. Another aspect to stewardship is recognizing when issues are pervasive or recurring and looking for opportunities to solve the larger issues instead of each individual instance.

- *Leadership.* There are several leadership actions that need to be taken during the issue management process. Program managers should look to resolve the issues as quickly as possible; this may require negotiating with other areas or presenting alternatives and a recommendation to stakeholders that will influence the decision and actions. Program managers will also have to demonstrate leadership when rallying and motivating the team around the issue and then driving it to closure.

4.9.7 Simplicity

A program can have many issues occurring across several projects at any given point in time so it is important to organize them logically and in such a way that anyone looking at the issue log can get the appropriate information. The same list of techniques suggested for simplicity with the risk log should be considered for the issue log as well. Programs may even want to consider housing issues and risks in the same place because of the synergies and tracking of similar data elements.

Although issues can be complicated in nature, it is important to organize and present them in such a way that the stakeholders can understand the issue background, effects, and options to be able to make an informed decision. Unorganized materials could confuse decision makers and have them spending more time focusing on understanding the complexity than on making the right decisions. Because sometimes team members tend to get into specific details and want to document all of them, the program manager should consider only including facts that support the objectives and use the rest as supporting materials.

4.9.8 Taking a Customer Approach

There are a few internal customers within the issue management process that need to be recognized and managed.

- *Affected areas.* There are two types of affected areas that are the recipients of the analysis and the decision regarding an issue, those areas affected by the issue and those affected by the decisions made. For example, a project has had a resource who was working on a critical deliverable leave the project. The project that he was working on is now at risk to meet its dates but then there is also an impact on the organization that has to supply a new resource to backfill the one that left.

- *Decision makers.* In order to make fact-based decisions on how to resolve an issue, the decision makers will need information regarding the issue and the quantified impacts. This information will need to be captured, organized, and presented in a way that allows these stakeholders to make optimal decisions. The program team should understand how these stakeholders prefer information presented and then tailor the approach accordingly.
- *End customers and stakeholders.* Issues can affect the end customers of the program, which can include other projects, sponsors, organizational units, or the end customer of the company's products and services. These stakeholders need to understand that there is an issue and what that means to the program commitments that were made to them.

4.10 DECISION MANAGEMENT

Decision management is a fundamental aspect of managing programs and yet it often does not get the same attention as other functions such as finance, schedule, or resource management. In fact sometimes we do not even recognize that we are within the decision-making process. Jim Nightingale, author of *Think Smart–Act Smart*, states that "We simply decide without thinking much about the decision process" (Nightingale 2007). It is important to be conscious of the decision-making process because of the implications for our programs. Figure 4.14 identifies the five primary steps in the decision management process:

1. *Define problem.* This initial step involves the identification of the problem that needs a decision and the desired goal. It is important to isolate the problem so that there is clarification of what decision is required.
2. *Develop options.* Once a problem is identified information related to the problem needs to be gathered so that options can be formulated. An option can also include not taking action.

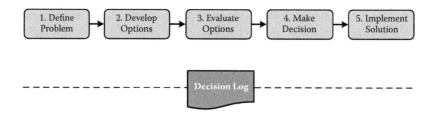

FIGURE 4.14
Decision management process.

3. *Evaluate options.* The options need to be evaluated with regard to how well they solve the problem as well as any consequences of implementing the option. A set of criteria should be considered when evaluating an option.
4. *Make decision.* After the options have been considered according to relevance, impact, and consequences, a decision should be made. A decision log is a useful tool to document key decisions and relevant information.
5. *Implement solution.* The recommendation that was decided on then needs to be implemented and monitored. In some cases the outcome also needs to be reported back to the decision makers.

Program decisions are required every day and they usually have significant implications on many aspects of the program. Table 4.11 identifies just a few examples of decisions that get made within each of the program

TABLE 4.11

Examples of Program Decisions

Function	Decision Examples
Work Intake	• When a new project can start based on available resource capacity
Scope and Change Management	• Whether to accept a change and the implications for the program
Schedule Management	• Action required for activities that are slipping with the risk of missing dates
Financial Management	• Actions on how to manage unplanned costs • What to do with projects that are forecast to be over budget
Resource Management	• What work to allocate resources to and who the best resources are to assign
Vendor Management	• Whether to extend an existing contract or look for cheaper alternatives
Issue and Risk Management	• Option selection regarding the mitigation of a risk or the closure of an issue and accepting the consequences

functions. Making sound decisions is imperative to the success of a program and there are many decisions that happen every day. The only way to make accurate and informed decisions is to follow the guiding principles throughout the decision-making process (starting with being conscious that decisions are being made).

4.10.1 Diligence

Program managers have to make decisions daily that have impacts across the projects and programs that they manage. Having diligence with regard to making sound decisions will allow these decisions to optimally affect the program and therefore increase the probability of success. Several examples of diligence are highlighted below:

- *Determining that a decision needs to be made.* Having diligence in the management of the program will provide the insight that a decision needs to be made. Without staying on top of each of the program functions, the need for a decision may get overlooked and therefore the effects of not making that decision will be increased. An example of this could be a resource who is assigned to two critical program tasks at the same time. Diligence in managing the schedule would enable the program management team to recognize this conflict and make the decision as to how to resource both activities. If the team was not managing the plan then they would realize this conflict once the work started and one of the activities would not get completed.
- *Performing the option analysis.* The identification and analysis of options is the key to making effective decisions, so the team needs to be thorough in their analysis. Being thorough includes using the right facts and logic to document each alternative and understanding the consequences of each option.
- *Involving the right people.* Because decisions can have impacts on many different areas and stakeholders, the right people need to be part of the process. This can include involving affected areas in the analysis as well as making sure that some stakeholders are even part of the decision-making process.
- *Considering a decision-making framework.* There are several techniques for utilizing a decision-making framework a program team may want to consider using as a structured way to consistently make decisions. The simplest approach is to list the positives (pros) and the

negatives (cons) for each option as a comparison. Other frameworks identify specific roles for decision makers, influencers, and contributors with each role having a specific function and set of responsibilities.

- *Documenting the decision.* Once a decision has been made the team should document that decision in a decision log which should include a description of the decision, list the decision makers involved, and document the justification. Documenting the decision will provide an historical reference for the program in case a similar decision is needed in the future.
- *Implementing the decision.* Once a decision is made, the program team will need to implement the decision with diligence to make sure that it does not cause additional problems or unplanned consequences.

4.10.2 Attention to Detail

Building upon the diligence needed for decision making is also monitoring the details related to those decisions. The details are where the specific implications get understood and managed and there are several considerations regarding this:

- *Focus on what is important.* There are usually many details and considerations related to a decision, so the program team should focus on only those details relevant to the decision.
- *Details of implications.* As with other program functions such as risks, issues, or changes, the team needs to consider all the detailed implications of the decision on scope, resources, cost, schedule, and quality. These details are important because they will factor into making the decision and will allow for the appropriate action to be taken when a decision does get made.
- *Decision log management.* Paying attention to details is needed when managing the decision log because that is where all of the key information is stored. This includes making sure that the key points are documented and any follow-ups get identified with assigned dates and owners.

4.10.3 Transparency

Because so many decisions get made on a program that have implications there needs to be transparency across the entire decision-making process.

For starters, transparency into the tracking of program information is what is used to identify that a decision is needed. The early identification of a situation that requires a decision allows for time to make an informed decision and act. Without this early indication, the decision may have to be rushed without all information and therefore become risky. Having transparency of program information also provides key datapoints for an accurate depiction of the implications so that an informed decision can be made.

Transparency is also used in the documentation of the decisions in a central decision log. The decision log then becomes the single source of truth for all decisions and the place for team members to go to understand the history of the decisions. Oftentimes and especially on big programs the teams find themselves questioning prior decisions or direction and having the historical records of the decisions becomes useful and important to understand. Having these decisions documented with the specific reasoning and assumptions also prevents confusion as well as conflicting opinions regarding the historical recollection of the decision.

4.10.4 Single Sources of Truth

Within the decision management function, a decision log is used as the central repository of all decision information. This is important to have because many times team members have different recollections and interpretations of the decisions or they weren't involved and want to understand the rationale for them. Also it is hard to manage decisions that are buried within meeting minutes or e-mail threads as opposed to having them in one central location that the team can reference.

There are several key decision elements that should be captured and tracked in the program decision log. As with the other sources of truth, these elements need to be maintained and updated as information becomes available:

- Date that the decision was made
- Project or work stream affected
- Description of the decision that was required
- Description of what the decision made was
- Identification of which stakeholders made the decision
- The rationale and assumptions used to make the decision
- Any follow-up actions required with dates and names

As decisions get made, the program team also needs to remember to update other single sources of truth if the decision has an impact on them. For example, a decision made on a resource could require changes to the cost forecast or schedule milestone and therefore those documents would need to reflect the changes. The checklist used to update documents after changes are made could also be used here.

4.10.5 Fact-Based Decisions

At the heart of the decision-making process is using facts to make an informed decision. Facts allow for a more accurate decision that has predictable and manageable consequences. There are several considerations for using this guiding principle.

- *Focus on necessary information.* There are many facts that can be used when making a decision but the team should consider only those facts that are relevant to the options and consequences. Having too many facts may confuse the stakeholders or divert attention to irrelevant topics.
- *Confirm the validity of the facts.* Inasmuch as decisions have consequences it is important to have confidence in the facts that are being used to make the decisions and state when the information used may not be fully accurate. For example, financial implications may be high-level estimates so the assumptions used should be documented and shared with the decision makers.
- *Use a set of criteria.* Options to resolve a problem should be measured against a common set of criteria so that there is a balanced comparison between the alternatives. Section 3.6.1 identified a method called quality function deployment (QFD) whereby options can be scored against a weighted set of criteria for a quantitative analysis. These criteria could include cost, schedule, quality, or scope and should be elements that are important to the stakeholders who are making the decision.
- *Define decision makers.* Program managers need to ensure clarity around which stakeholders provide input, review recommendations, and make decisions. This should be clear as the decision-making process unfolds so people are aware of their roles and expectations.

4.10.6 The Ships

Program managers must make many decisions related to their program as well as influence many other decisions. In order to ensure that the best decisions are made they need to have the mind-sets that are enabled by the three ships.

- *Ownership.* This means owning the need to make a decision based on facts as well as the consequences of that decision. Ownership means not just ensuring that the decision gets made but rather having a feeling of accountability over the options and driving for the best possible outcomes for the program and the company.
- *Stewardship.* Making effective decisions involves a sense of stewardship of the program and the company. Most company strategies are realized one decision at a time and program managers need to recognize the implications of their decisions in relation to the objectives of the program as well as the strategic direction of the company. For example, many program teams have choices to make tactical decisions that may meet immediate objectives of schedule or cost but may not be in alignment with a company strategy. These options need to be recognized and then weighed so that the best decision is made for long-term and short-term objectives.
- *Leadership.* Program managers need to demonstrate leadership throughout the decision-making process by either making sound decisions themselves or by influencing other stakeholders to do so. Leadership requires persuasion, communication skills, and facilitation which are all characteristics of a good leader.

4.10.7 Simplicity

Programs today are extremely complex and have many parts associated with them. However, in order for sound decisions to be made, the concept of simplicity needs to be used. This does not mean ignoring all of the details but rather focusing on the right details and then presenting them in a way that is easy to understand by the decision makers.

As described in the fact-based decision subsection of this chapter, the program team should focus on facts that are relevant to the decision. This approach will avoid some of the noise associated with having too many details and will allow the team to home in on the relevant information. Also if the team concentrates on the specific problem to be solved, then the discussion can be very specific and avoid the tendency to address the events leading up to the current problem.

Once the information is gathered it will need to be organized in a way that is easily interpreted and meaningful to the decision makers. This could include quantifying the impacts of each option, using visual representations of impacts, or a side-by-side comparison of the options across common dimensions.

Lastly, the decision-making process should also embody simplicity. A decision-making process that is unorganized or that has many steps will result in confusion and slow progress. The decision-making process should lay out the roles and process in a way that promotes clarity and efficiency. For example, there should only be one decision maker identified for each decision.

4.10.8 Taking a Customer Approach

There are a few customers within the decision making process:

- *Decision makers.* Decision makers need information regarding the alternatives and the quantified impacts so they can make informed decisions. As noted in the simplicity section, this information will need to be captured, organized, and presented in a way that allows these stakeholders to make optimal decisions. Also it is important to understand how these stakeholders prefer information to be presented so they can organize the relevant documentation to meet their needs.
- *Affected stakeholders.* There are program stakeholders who will be affected by the decisions and are therefore customers of the decision itself. They need to be involved in the process and made aware of the decisions made and any actions they are required to take as a result.
- *Program team members.* Decisions result in taking action and the program team members who are responsible for these actions need to understand the specifics of the decision and of the actions now required of them.

4.11 ACTION ITEM MANAGEMENT

Managing large programs means that there is a significant amount of work to perform by the program team. The key program activities and deliverables should be tracked within the project schedule but there are also many other tasks that go beyond the planned activities which need to be tracked. These additional tasks could include preparing for important upcoming meetings, estimating a new piece of work, acquiring specific resources, or simply the takeaways from project and program meetings. All of these tasks need to be tracked and managed and are done so through the action item management process. The steps in this process are outlined in Figure 4.15.

1. *Identify actions.* The first step is to identify the actions that need to be tracked. The most common source of action items is from discussions held in meetings where follow-ups are requested and agreed upon. Actions can also come from other areas such as thinking through planning activities or simple reminders of tasks that are coming up.
2. *Take action.* Once it is identified that action is needed, the specific steps need to be determined and tracked. A central action item log can be used to document the actions, owners, due dates, and progress.
3. *Follow up.* Lastly, after the actions are completed there should be a follow-up to confirm completion or any additional actions that may be needed. Some action items will only require being checked off when they have been completed, whereas other action items may be recurring or may require additional follow-up.

With lots of activities that arise during the planning and execution of a program it becomes essential to identify and then track all of the action

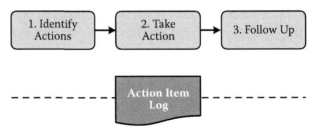

FIGURE 4.15
Action item management process.

items diligently and in one place. Without using the guiding principles to manage actions they will become disorganized, be forgotten, or not complete when required.

4.11.1 Diligence

The management of action items requires diligence to identify, track, and then follow up on the specific actions. There are a few considerations that should be used when managing program action items.

- *Identifying actions.* Oftentimes in meetings there are several action items that get identified and the program team needs to make sure they get captured. This may mean the identification of a scribe before the meeting so it becomes clear who should be taking down the actions and assignment of owners.
- *Assignment of one owner.* Each action item should have one owner assigned to it, which provides accountability and avoids the scenarios where two people are assigned and both think that the other is working on it. This approach also allows the program manager to know who to follow up with for closure of the action. It is also a good idea to assign ownership to a person who is part of the team or meeting where the action item gets assigned; otherwise it may be difficult to ensure that the owner is making progress.
- *Monitoring dates.* As with managing schedule commitments, the program manager needs to monitor the dates that the action items are due and ensure they are completed on time. Team members have many activities to perform and sometimes may need reminders of these actions and their expected completion dates. Having the actions in one place and reviewing them frequently allows for these proactive reminders to occur.
- *Keeping information updated.* As with the other program functions, the information regarding action items needs to be current and accurate. Without this diligence the action item log will become just another list that no one is using in which case there is no point in tracking actions anyway.
- *Recognizing an issue.* Team members should be careful not to track issues in the action item log and to recognize when something needs to be managed separately in the risk or issue management process.

4.11.2 Attention to Detail

Not only is diligence required to manage the action items, but the team also needs to pay attention to the details of the actions. An unorganized or outdated action item log will result in oversight of activities, missed target dates, and a lack of clarity for program team members. There are some specific areas to pay attention to regarding action items and they include the following examples:

- *Recognize actions.* The team needs to be conscious of when a takeaway is identified to make sure that it gets captured on the log. Sometimes meetings can jump around between topics so it is important to recognize when there are actions so they can be captured properly.
- *Identify accurate actions.* The identified actions should be specific and accurate which may require confirming them with the action owners. If the note taker in the meeting is not familiar with the content then she may not have captured the action description entirely correctly so it should be validated.
- *Monitor dates.* The action item log should be reviewed on a regular basis with special attention to closure dates and status. This technique allows for follow-ups if there are many action items that are past their target completion dates or dates that are getting close without significant progress.
- *Monitor progress.* The program manager needs to pay attention to how the work is progressing to expected dates and track this progress in the action item log so that there is an historical set of comments.

4.11.3 Transparency

The principle of transparency is used in the identification of program action items. By having detailed tracking and reporting across all program management functions, the program manager can identify where additional action items are required. For example, monitoring the program resource roster could reveal that additional resources are required and that the team will have to put in a requisition for hiring a new position (most companies have multiple process steps and forms to fill out). Each of these steps should be captured and then tracked in the action item log.

Also, having a master action item log provides transparency into all of the tasks that need to be performed beyond the activities in the program schedule. This technique is important because there can be many of these little tasks which could add up or be forgotten as the primary program work gets performed. By having the information captured in one central location and managing it to be constantly reviewed, the team members are aware of these actions and the expectations to close them.

4.11.4 Single Sources of Truth

A centralized action item log is needed to capture all of the project and program actions in one place. Often, these actions can get buried in meeting minutes (if captured at all) and are not aggregated in one place. Therefore, a program team member would have to sort through several meeting documents or e-mails to understand the historical actions and assignments instead of having one central place to see them all. There are several core elements that should be captured in the action item log:

- *Action item number.* Assignment of an identification number for each action item.
- *Date created.* The date that the action was identified and documented.
- *Action description.* A description of the action being tracked.
- *Category.* A category for the action item could be the project name, organization, or meeting to which the action relates.
- *Assigned to.* The name of the program team member who is accountable for completing the action item.
- *Due date.* The date that the action item is targeted to complete.
- *Status.* A status of the action item which could include being open, in progress, closed, or deferred.
- *Comments.* Comments and notes regarding the progress of the action item toward the date. These notes should have dates assigned to them so there is an historical record of progress.

4.11.5 Fact-Based Decisions

Although action items may not require decisions directly, they may be used to gather the facts needed for a particular decision. As decisions are identified they usually require several pieces of information to be gathered and analyzed. These steps to prepare for a decision could be tracked as specific action items. For example, to compare two options in order to make a decision there may be action items to follow up with specific team members of organizations to gather information that can then be used in the analysis.

Also the principle of gathering facts can be applied to tracking action items. This can include having an accurate description of the action as well as the updated history of progress against closing the action. Having the historical inventory of actions and progress against them could need to be revisited during the program. For example, if there is a question regarding how the action was handled or when it was closed those facts would be available from the action item log.

4.11.6 The Ships

Diligently managing the many actions that arise during the course of a program requires using all three of the ships.

- *Ownership.* The person who is assigned the action needs to take it on with a sense of accountability to work it and drive it to completion. This should mean being self-motivated to meet the deadlines and that she should not require someone else to follow up with her to do so.
- *Stewardship.* Many of the action items can have impacts on the program commitments of resources, schedule, or cost. Therefore, being diligent with completing the action items means being a good steward of the program and the company.
- *Leadership.* Signing up to own an action item is a demonstration of leadership especially when there are actions that no one wants to step forward to undertake. Completing the action item may also require leadership traits such as influence, perseverance, negotiation, and facilitation.

4.11.7 Simplicity

In the case of action item management, simplicity can mean having an easy-to-use and understand action item log. The actions should be organized and presented so that it is easy to navigate for stakeholders. This can also include filters and grouping the action items especially as the log gets filled with open action items and can start to become large in size. Also structuring the work so that there is clear ownership and dates is important so there is an easy way to acknowledge and track the work.

If the action item log is used during meetings, the program manager may want to consider a simplified version of the actions to present. This could include filtering the action items so that only the ones relevant to that meeting are shown. This way the information does not get overwhelming or confusing for team members and the focus can be on the specific actions that are relevant to that meeting and audience.

4.11.8 Taking a Customer Approach

The action item management process can have several internal customer relationships that need to be managed:

- *Action owner.* The action item owner is accountable for the action item but may also be the customer of the activities required to close that action item.
- *Affected stakeholders.* The actions taken may have implications on other areas of the program or on other stakeholders. These stakeholders should be aware of the actions and possible implications.
- *Program manager.* The program manager needs to track all of these action items so they are a customer of the status against completing the actions.

4.12 COMMUNICATIONS

Communication plays a significant role in program management. By some industry estimates program managers or project managers spend 80 to 90% of their time involved in communications of some sort. For a program manager to be successful he must use several types of communication which involves constant transparency of relevant information. Program communication is not only about producing status reports and management presentations, but is also about getting the right information to the right people at the right time. There are several examples of information needs for a typical program:

- *Status reporting.* Stakeholders need to understand the progress of a program related to its commitments. This includes program health, tracking toward financial targets, deliverables completed, upcoming milestones, and any issues or risks that require attention.

- *Management reporting.* Management needs to know progress toward goals and any escalation of issues, risks, or changes that can have an impact on the program commitments. These need to be organized and documented properly with a description of the challenge, impacts, and alternatives with a recommendation.
- *Stakeholder meetings.* Large programs have many stakeholders with different agendas and interest in the program and therefore several meetings are required to keep them informed or even involve them in the management or governance process.
- *Program team member understanding.* Team members can tend to get focused on their specific activities and then lose focus on the big picture. The program manager continually needs to communicate the vision, priorities, approach, and plan to the team members to keep them focused and at the same time aware of the overall progress and goals. They also need to understand the day-to-day health of the projects and how the work they are performing is trending.

There are the three key activities involved in program communications, including communication planning, collecting information, and then presenting information. These activities are shown in Figure 4.16 and should be considered iterative in nature and not one-time activities.

1. *Plan communications.* As programs become more complex, they have many moving parts and stakeholders that require information to help the program to meet its commitments. A program manager needs to spend time on stakeholder analysis and communication

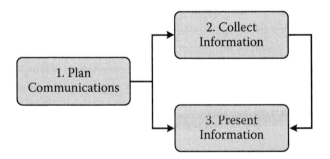

FIGURE 4.16
Communication activities.

planning to have a solid approach for managing the information between these moving parts. These should be documented in a communication plan.

2. *Collect information.* A program team needs to be constantly gathering relevant and timely information that can be used to communicate to stakeholders. This can include progress regarding the program health, facts needed to make decisions, or information on the program background and goals.

3. *Present information.* As the information gets identified and collected, it needs to be presented to the appropriate stakeholders. Methods to present the information can include recurring status reports, stakeholder meetings, or even informal interactions with team members.

Because of the importance of communications on the success of a program, the program manager needs to utilize all of the guiding principles. Having a solid communications plan and executing it well could mean the difference between a successful program with involved stakeholders or confusion and delays.

4.12.1 Diligence

Diligence should be taken when creating the program communications plan to ensure that the right stakeholders are identified and that the plan is comprehensive and meets their needs for information. This plan should be thorough because stakeholders who are not involved or do not get the information they need could cause problems for a program such as political noise or conflicting decisions. Once the plan is created, the team then needs to be diligent in the execution of that plan.

Diligence is also required when gathering the information that will be presented. A program team needs to collect the appropriate information from several different aspects of the program related to its health.

- *Strategic goals and scope.* Most program communications should somehow be grounded back into the program goals and scope. This approach sets the context and reminds people why the program exists.
- *Progress toward goals.* This includes status of activities, progress toward key milestones, and starting activities when scheduled.

- *Affecting items.* The program needs to identify issues, risks, and changes as well as any associated effects on the program's goals. Once identified, appropriate information needs to be gathered to determine the magnitude of the impact and assess any recommendations or decisions that need to be made by management.
- *Financial information.* This information includes data on actual spend-to-date and forecast spend remaining. This usually involves some financial analysis to determine trending and forecasting.
- *Risks to delivery.* Risks can come from any aspect of the program and the program manager needs to understand where they are and be able to monitor them closely so they do not evolve into major problems on the project.
- *Upcoming decisions.* This information can be used to manage expectations around any upcoming decisions that stakeholders will need to make.
- *Realization of benefits.* The program should track information regarding any expected benefits and how the team is progressing towards those.

4.12.2 Attention to Detail

Because there are so many types of communications on a program that are going to many different stakeholders, paying attention to details is required. First, it is important that the specific details documented in the communication are accurate. If these details are not correct then either the stakeholders will be getting wrong information or the credibility of the program team will be diminished. Datapoints in communications should be validated and reviewed with key stakeholders prior to being distributed.

The other area to pay attention to is the formatting and structure of the communication. Although the content of a communication is the most important component, the format should also demonstrate high quality. Presentations that have inconsistent fonts, unstructured text, and pictures that are not aligned appear sloppy, which may take away from the credibility of the content. A presentation should be seen as a reflection of the quality that a program manager expects of the team and his or her own work. There are several ways that a communication can have high quality and attention to detail:

- Focus on telling the right story and then using the relevant details to substantiate the key points.
- Colors are an effective method of highlighting critical items and bringing attention to risks and issues.
- Visuals and diagrams can be used to highlight key messages and are more appealing than reading long sentences. They also break up presentations that are all text where key points can be missed because they are buried in paragraphs.
- Focus on the formatting such as making sure that bullets are aligned consistently, words wrap properly, and that images are lined up.
- Use categories to group key messages instead of having many long sentences all blended together.
- Do not fill the communication with many words and sentences but rather keep it to relevant information.
- Make sure that names are spelled correctly and correct spelling and punctuation are used. This sounds obvious but I have seen many times where this is not the case (and those were presentations to senior people).

4.12.3 Transparency

The essence of program communication is transparency of information to stakeholders. By managing work diligently and then having ample communication, the team can realize many benefits:

- Updated and relevant information regarding the program status, structure, milestones, risks, and issues. This technique allows the entire team to be informed of the progress and be on the same page regarding progress toward program commitments.
- An understanding of key actions and decisions that are required and the facts supporting the options and recommendations.
- Common expectations of priorities, strategic objectives, scope, and commitments of the program.
- Engaged and informed stakeholders who can champion the program and help make difficult decisions.
- A sense of order and control over the program information.

TABLE 4.12

Communications and Transparency

Communication Vehicle	Transparency Provided
Status report	• Health of projects and programs along several criteria (e.g., schedule, quality, financials, resources) • Progress against interim and end milestones • Description of key issues and risks with actions required to address them
Project and program meetings	• Discussion of progress, dependencies, changes, issues, and risks • Key program information including priorities, upcoming milestones, and important messages • Progress metrics such as defect counts
Stakeholder meetings (e.g., steering committee)	• Progress against goals • Items for management attention such as changes, issues, or required decisions
Program newsletter	• Key messages for the program team which can include standards, upcoming dates, or team accomplishments • Consistency of messages for the entire team
Visual displays such as posters or dashboards	• Having key progress indicators for the team to see in the workplace
Program calendar	• Tracks key recurring meetings with relevant information

There are several ways that program communications can be used to provide transparency of information. Table 4.12 identifies some examples of commonly used program communication vehicles along with the transparency of information that is being provided to the program stakeholders.

Lastly, all program communications should be stored in a common file share so that program team members can find any historical information regarding key stakeholder communications or decisions. It is a good idea to put the date of the communication in the file name or on the communication itself so when people look for the documents it is clear when each communication occurred.

4.12.4 Single Sources of Truth

Creating a master communication plan is an important component of program success. Ineffective or inconsistent communication can result in delayed decisions, rework due to stakeholders not getting information early, differences in expectations, low morale, and missed commitments. A communication plan should be created early in the program and then managed throughout the life of the program. There are a few fundamental elements that should be documented in the communication plan:

- *Stakeholder identification.* Create an inventory of the stakeholders who require communication as well as what information they need. Stakeholders can include team members, management, other divisions, managers of program resources, and vendors. Each stakeholder may require different information so it is important to consider and document each one.
- *Format.* Identify the format of communications which can include meetings, status reports, newsletters, or e-mails.
- *Frequency.* Determine the frequency of each communication (e.g., a weekly status report or monthly steering committee meeting).
- *Source.* Identification of the source that will be used to provide the information for each communication.
- *Owner.* Naming an owner from the program who will be responsible for each communication vehicle.
- *Comments.* Any relevant comments or notes regarding the communication approach or stakeholders.

Also, having several known sources of truth for the program enables better communication because there is clarity over where the key information can be found. This information can then be used to create the communications effectively instead of trying to start over each time a communication is needed. For example, all program schedule information should come from the master schedule and all financial information should come from the financial reporting.

4.12.5 Fact-Based Decisions

In order for program communications to be effective, the facts used need to be accurate and timely. These facts can be used to represent the status of the program, to highlight the effects of an issue, or to facilitate a decision that needs to be made by management. Also, by using facts in communications it takes away blame, opinions, and politics. By using the principles of diligence, attention to detail, and single sources of truth a program should be able to have the right facts available for credible communications.

Another way to gather or validate information for communications is a technique called management by walking around (MBWA). This is an approach where the program manager walks around to the team members and has informal conversations with them on work-related topics (or calls them for virtual teams). MBWA is a great technique for genuinely interacting with the team, establishing their trust, and getting qualitative program information. This technique can be used to validate information in a communication, clarify facts, or simply get the opinion of the team member. Beyond getting information, this is a very effective technique at building rapport and trust with the team members.

4.12.6 The Ships

Communications is at the core of both leadership and program management and therefore the ships become important principles to use:

- *Ownership.* Communications need to be made from a mind-set of ownership and not simply reporting the status. Taking ownership of the communications means having accurate facts, being able to explain the key messages, and utilizing the information to facilitate decisions or action. Ownership also means delivering on the communications in the plan to which the team member's name is assigned.
- *Stewardship.* When program managers communicate on how the program is progressing toward meeting its goals they are demonstrating stewardship. Program managers are also the stewards of the key messages and information and utilize communication techniques to share those with stakeholders.
- *Leadership.* Program managers need to create an environment of trust for team members to share information. This approach demonstrates leadership for the team members and also allows for credible and accurate information to be presented. Leadership also means using communication methods to deliver difficult news or facilitate complicated program decisions.

4.12.7 Simplicity

Simplicity is a very important concept with regard to program communications. This starts with recognizing and regarding the audience of the communication. A program manager needs to consider who the presentation will be given to which should then influence how the presentation is created. For example, some executives like the end results presented first with the supporting data as backup and others like the information to tell a story that "builds up" to the end results. Also, some managers like to see data that support the findings and some like to see abstract diagrams. These preferences should be documented in the communication plan and then applied during the communications.

Once the audience is understood, the information should be presented with only relevant facts and in an easy-to-understand format. Many presentations become overwhelming with full paragraphs of text or tons of data which can be confusing and take the focus away from the key messages. Presentations should tell a story in a simple manner and use facts or diagrams to substantiate the key points.

Another relevant consideration of simplicity can be the number of meetings that a program has. Often the amount of meetings that occur on a program can easily become overwhelming as can the number of participants in those meetings. Program teams should take regular checkpoints to consider the amount of meetings and participants to rationalize and optimize them to make the team more productive.

4.12.8 Taking a Customer Approach

Inherent in any program communication is the customer of the information who is in the target audience. First and foremost, any communication needs to be planned in such a way as to provide the recipients (i.e., customers) with the type of information important to them in a way that makes sense to them. Presentations are not about what you want to say as much as about what type of information the audience is looking to hear and how they want to hear it. This is why the communications plan should be considered an investment in time and especially the planning around stakeholder identification and analysis.

4.13 PROGRAM OPERATIONS

The last program management function to be described in this book is the bucket of everything else that is called "program operations." This function includes all the activities that keep the program moving and support the other functions. Table 4.13 identifies several of these program operation functions.

TABLE 4.13

Program Operation Functions

Function	Description	Examples
On-Boarding	Orienting new program team members to the program goals and processes	• On-boarding checklist • Orientation meeting and materials
Off-Boarding	Updating key information when a resource leaves the project	• Off-boarding checklist
Program Standards	Documentation of key program functions, processes, policies, and tools	• Procedure manual • Standards and templates • Travel policies
Logistics	Management of space including meeting rooms and seats	• Conference room calendar • Seating chart • Conference call numbers • Shared team calendar
Document Management	Central location for storing program documents and standards	• Program file site • Document and meeting standards

These items may seem of little impact but they are the "glue" that holds everything together on the program. If team members were not brought on successfully or conference rooms were not available, there would be delays in program work. Similarly without document management standards or program processes, the program materials would quickly become disorganized and quality would suffer. The guiding principles should be applied here as well to ensure the proper level of quality on the operational functions.

4.13.1 Diligence

Although the program operational items may be considered administrative functions they still require diligence in their execution. There are several examples where diligence should be used for each of the program operational functions:

- *On-boarding/off-boarding.* Having a checklist for bringing on and rolling off resources will enable the team to perform each step. This can include things such as providing resources with access to systems or computers that will enable them to be effective when they start the program.

- *Standards.* Program teams should have all of their processes documented in one common place that explain how the program should function. Note that these processes may be in the context of existing organizational processes. By having these standards documented the team can then operate using them properly and avoid missing key activities that could require rework.
- *Logistics.* Ensuring that team members have a place to work in the office and that meetings have appropriate conference rooms allows the team to be productive toward goals. Not having the right logistics can result in delays of critical conversations or unproductive work time.
- *Document management.* Programs produce a tremendous amount of documents and by having diligence in the structure and planning of files a program can control and organize these files. Otherwise the documents start piling up all over the place and there is no clarity over where to find anything.

4.13.2 Attention to Detail

The program operations function also needs to pay attention to the details given the number of activities that it supports. For example, the logistics function requires many details that need to be managed. Because programs require many meetings when there is a problem with logistics such as not having a room to meet in or a conference call number to dial into, the result may be a lack of productivity and delays of critical decisions.

Another area to pay attention to would be confirming that program procedures are accurate and updated. If the expectation is that program team members follow these standards, then they should be correct. There are several areas to pay attention to with regard to standards and processes:

- Having a complete set of standards that are modified based on any feedback or changes in the company standards.
- If the standards have any calculations in them they need to work properly.
- Confirm that any links to key sites or documents are working. Sometimes documents can get moved or renamed and then the links no longer work.
- Keeping a revision history of documents to capture any changes.
- Providing examples and templates that team members can use as starting points.

4.13.3 Transparency

By documenting program procedures and standards the team members will have transparency into the expectations, policies, and standards for operating within the program. This technique will allow for a consistent approach, execution of work, and use of templates.

Another form of transparency is through the management of program logistics. By having a shared team calendar and meeting calendar, team members can see when rooms are available and when key team members may be out of the office. This insight will allow team members to plan around certain events or absences instead of finding out about them when it is too late and there may be implications for existing schedules or meetings.

Also, having an organized structure for program documents is an important way for team members to find key deliverables and historical artifacts. Many times, companies that have standard methodologies also have standards for document storage and naming. Absent that, a good technique is to organize the documents in folders that align with the projects and phases. The number of folders should also be kept to a minimum so information can be found easily. Often, program document sites get cluttered with files and it becomes hard to tell where anything is. Program managers should consider having a role on the team to be the project librarian to manage the documents. It is a small amount of effort to keep the library clean but well worth it for the program team.

4.13.4 Single Sources of Truth

There are several program operations documents that should be used as authoritative sources of information:

- *On-boarding/off-boarding checklist.* Most companies and programs have many activities that need to be completed for new team members and can include system access, seat designation, obtaining a computer, addition to distribution lists, and invitations to key meetings. Similarly as resources roll off the team they need to be removed from distribution lists and meeting invitations. Program teams should create central checklists to identify and then track all of these items.
- *Standards.* Program teams should also have their standards and templates documented in one place which will enable team members

to be able to easily understand the policies. Having these documents in one place will also support using the standards more than if they are hard to find.

- *Logistics.* There should be a common place for the team to go to find logistical information such as a shared team calendar, a conference room schedule, conference call information, or travel policies.
- *Document management.* Program documents need to be located in a central program file structure and organized well for the team and other stakeholders to be able to find key documents easily. It may even make sense to create a list of key documents and links for the team.

4.13.5 Fact-Based Decisions

Having structured program operations and logistics can help enable the team to make fact-based decisions. Several examples of this are listed below:

- Providing available meeting rooms and conference call numbers so the appropriate planning and decision meetings can occur.
- Documenting the processes to gather the facts as well as the framework for making the decisions.
- A centralized and organized document structure can enable decisions through helping to make it easier to find key documents and facts.
- If decisions require adding or removing team members this can be done effectively by using the proper checklist.

4.13.6 The Ships

Although program operations may seem to be of low importance to a program relative to some of the other functions, the program manager still has to manage them in the same way as the other functions within the program which includes using the ships:

- *Ownership.* Having ownership over the program also means taking accountability for all aspects of the program including orientation of new members, having the right logistics, and managing expectations through documented standards and processes.
- *Stewardship.* Having a structured document management process will provide stewardship for the key deliverables and information

for the program. This is especially helpful for stakeholders who are not familiar with the program or as an historical reference once the program has ended.

- *Leadership.* There are many opportunities to demonstrate leadership with the program operations. One example could be for the program manager to attend the orientation session as a way of setting the tone and direction for the new members of the program team. Also by showing diligence in all aspects of the program the program manager is setting the expectations for how to operate in the program.

4.13.7 Simplicity

Program operations can get overwhelming if not managed properly especially on large program teams that have many meetings and resources coming and going. Simplicity should be considered when setting up the operational processes:

- *On-boarding/off-boarding.* Keeping all of the activities in one checklist should make it easy for resources to understand what they need to do when they join the program. This is a simple approach and yet many programs neglect using one and rather have a "throw them into the fire" approach.
- *Standards.* It is important to document processes, procedures, and standards but the program team should be sensitive to not over-complicating these. Although having a 100-page manual may be comprehensive it also pretty much ensures that no one will ever read it. Materials should describe the key points but then show them in a way that is easy to understand and use.
- *Logistics.* Similar to the approach to standards, logistics should also be easy to understand using simple explanations and processes. For example, program teams should have one clear place to go to find out logistical information and an easy way to request a conference room or on-board a new resource.
- *Document management.* Simplicity should definitely be considered when storing program documents as file shares can quickly get out of control with many unstructured documents and folders. The folder structure should be consistent but also simplified to only those mandatory folders.

4.13.8 Taking a Customer Approach

As with every function in program management, there is a customer relationship within program operations as well:

- *On-boarding/off-boarding.* The resources who are joining or leaving the programs are customers of these processes. New program team members are customers of the orientation and on-boarding processes to try to learn as much information as possible so that they can get up to speed and perform their jobs.
- *Processes.* The program processes should be developed in such a way as to be useful for the program team members, who are the customers of this information. This can include having templates to leverage or simplicity in how the processes are described. The procedures and standards should also document the customer relationships for the key deliverables within the program team. For example, it can document that the design team is a customer of the requirement documents from the business analysis team or that the tester is a customer of the code from the development team.
- *Logistics.* Most logistics are done on behalf of program team resources so these people then become the customers of the operational process. Setting up an easy-to-use request and tracking system for logistical requests can be an easy way to provide value to these customers.
- *Document management.* All program team members and stakeholders are customers of the document management process because they all have needs to find and store program information. Setting up the document structure in such a way that it makes sense to the stakeholders and they can find what they need is a form of customer focus.

5

Conclusions

5.1 SUMMARY OF FUNCTIONS OF GUIDING PRINCIPLES

Chapter 4 went through each of the program functions and then aligned them to the application of the guiding principles. Figure 5.1 provides a high-level summary of the complete mapping of program functions to each of the guiding principles and can be used as a reference sheet for this book.

In looking at Figure 5.1 there are many common elements that are identified and consistent across the functions. The first observation is that each of the guiding principles has application across every aspect of the program. This can be used as a validation that program managers need to use the guiding principles pervasively across their programs. This is also consistent with many of the messages from the case studies used in the book.

The remainder of the book provides some final insights and themes around the use of the guiding principles. It is hoped that the case has been made for why the guiding principles are important and ample techniques have been identified to highlight how to apply them on programs.

5.2 COMMON THEMES

In this book, the eight guiding principles of the consultative approach have been identified and then applied to each of the program management functions. Examples and cases have been used to demonstrate the benefits

Function	Diligence	Detail	Transparency	Truth	Fact Decision	Ships	Simple	Customer
Work Intake	• Manage Pipeline • Analysis • Strategic Alignment	• Request • Accuracy • Estimates • Initiation Checklist	• Pipeline Report • Progress	• Pipeline Report	• Estimates	• Governance • Prioritization of Work	• Request Form • Pipeline Format	• Easy Process • Accurate Estimates
Schedule	• Thorough Planning • Milestone Tracking	• Accurate Activities • Dependency	• Monitor and Share Progress	• Master Schedule	• Dates • Durations • Dependency	• Activity Owners • Steward of Work	• Organized and Clean Plan	• Ability to Meet Commit
Finance	• Estimates • Actual Costs • Forecasts	• Math • Variances	• Trending • Overall Forecast	• Financial Tracking	• Actual Costs • Forecast • Variances	• Company Money	• Organized Finances	• Specific Views of Finances
Resource	• Align to Work • Skill Needs	• Allocations • Start Dates	• Open Roles • Alignment	• Resource Roster	• Resource Allocation	• Roles & Resp. • Growth	• Reporting	• Upcoming Needs
Vendor	• Assessment • Planning • Managing	• Quality of Work • Contracts	• Performance • Invoices	• Vendors • Inventory • Invoices	• Select Vendor	• Clear Roles • Mentor Team	• Process • Reporting	• SLAs • Reporting Needs
Change	• Assessment • Implement	• Accuracy • Updates	• Changes • Impacts	• Change Log	• Change Decision	• Manage All • Implement	• Tracking • Reporting	• Present Info • Reporting
Risk	• Identify • Assess/Act	• Understand Details	• Risks • Impacts	• Risk Log	• Determine Action	• Proactively Manage	• Reporting	• Share Impacts
Issue	• Identify • Assess/Act	• Understand Details	• Issues • Impacts	• Issue Log	• Determine Action	• Proactively Manage • Escalate	• Reporting	• Share Impacts
Decision	• Option Analysis	• Detail Implications	• Decisions • History	• Decision Log	• Make Decisions	• Make Best Decision	• Tracking	• Involvement
Actions	• Manage	• Monitor Closely	• Actions	• Action Log	• Determine Actions	• Owners	• Tracking	• Awareness
Comms	• Planning	• Details	• Information	• Comm Plan	• Agenda	• Own Messages	• Meetings	• Audience
Ops	• Management	• Accuracy • Quality	• Processes	• Processes • Onboard	• Enablers	• Documents	• Process	• Support

FIGURE 5.1

Program functions aligned to guiding principles.

of the guiding principles and the results of ignoring them. There are several consistent themes that are pervasive throughout the book and are summarized in this chapter. Many of these themes utilize multiple combinations of the guiding principles so the program team should recognize that it is the application of all of them that will produce the best results.

5.2.1 Constant Diligence and Attention to Detail

A program manager needs to expect diligence and attention to detail in everything from the biggest program deliverable to the smallest administrative activity. Solid planning without diligent execution is wasteful and ineffective. For example, the program team could spend a significant amount of time creating a perfectly accurate program estimate but then not managing change controls or issues properly may cause the forecast to stray significantly from the initial estimate. There are several areas where a program manager needs to stay diligent with a focus on the details and quality:

- *Tracking all requests properly.* Starting with the work intake function the program needs to demonstrate diligence in capturing, assessing, and tracking all requests for work on the program.
- *Managing the program and deliverables.* The program manager needs to understand the state of the program activities, resources, vendors, and finances at all times to understand progress and be able to communicate to all stakeholders effectively. Also being diligent in these functions will enable the program team to see trends and be able to take action proactively to avoid negative consequences.
- *Acting on changes, issues, and risks.* Identifying changes, issues, and risks and being thorough in the assessment of impacts so they can be acted on to allow the team to remain productive and focused.
- *Making decisions in a timely manner.* Decisions can hold up program commitments if they are outstanding for a long period of time and can also result in rework if not made properly. A program manager needs to be aware of what decisions need to be made, use the appropriate facts to facilitate that they get made, and then document them for future reference.
- *Tracking all action items.* As program action items get identified the team needs to document them with specific owners and target dates and then effectively manage them to their completion.

- *Continuing to gather and present accurate information.* Keep stakeholders informed of progress and key challenges and risks. This approach requires an understanding of the stakeholders' needs and thorough planning for proper communications.
- *Utilizing effective program operations.* This includes the on-boarding of resources, program logistics, and document management.
- *Using checklists for key deliverables.* This ensures that all activities get identified and completed. Three examples are listed below:
  - *Initiation.* A checklist for all projects coming into the program and the activities required to set them up
  - *Changes.* A checklist of all the single sources of truth that need to be updated after a change is approved
  - *On-boarding/off-boarding.* Checklists of activities and deliverables to update as resources come on and leave the program

With a diligent approach that focuses on details, a program manager can manage all of the functions listed above to help the program to continue progressing toward its commitments. It is crucial not to fall behind on any of these items, because then the program can quickly move into a state of being reactive to situations, causing activities to start piling up and become unmanageable. The creation of a program office is an effective way to help with the operational diligence and insights which can then allow the program manager to remain focused on removing obstacles and supporting the team. This approach is especially effective for larger programs that have many moving parts requiring management and insight.

5.2.2 Understanding Impacts

The second theme that is pervasive across the program management functions is to identify and understand impacts. This approach requires the guiding principles of diligence, attention to detail, and transparency of information to understand the facts and therefore the implications of the items listed below so that effective decisions can be made.

- *Intake requests.* New requests have to be understood in the context of how they align to business and technical strategies. They may be cases where a request for work is incongruent with a strategy and the team either needs to decide not to work on it or make a deliberate decision to stray from the strategy (with the right management participation in that decision). Requests also need to be estimated for impact on the program because decisions may have to be made based on financial and resource constraints as well as priorities of other program work that may have to change.

- *Estimates.* Any estimate has to be thorough and include all affected areas to understand financial, resource, and schedule needs. Because programs are complex most estimates involve people from many different organizations and other projects so program teams need to recognize these areas and include them in the process.
- *Forecasts.* The program manager needs to stay on top of understanding the financial and schedule forecasts that may have implications on the program commitments and require action to better align the forecast with the goals.
- *Vendors.* As program teams are assessing which vendors to use they need to consider the different criteria to analyze the vendor. Based on this analysis there may be effects of using the vendor that the team needs to recognize such as high costs or a lack of industry knowledge.
- *Changes.* Changes occur often on large programs and therefore they need to be assessed properly to understand the impacts on program characteristics such as scope, resources, cost, benefits, and schedule. Understanding these impacts will allow for the program team to make the right decision regarding the change and also update the program forecasts and deliverables properly.
- *Risks.* Program teams need to be diligent regarding the identification of risks and the analysis of each one to determine which require action based on their impact and probability of occurrence. Program teams also need to recognize what the impacts of not acting on the risk would be.
- *Issues.* Once issues occur they have an immediate impact on the program and so the team needs to thoroughly understand what the impacts are so they can be addressed properly. Note also that taking action (or not taking any action) on the issue may have implications and therefore these need to be assessed and recognized as well.
- *Decisions.* Making decisions requires choosing an alternative and accepting the implications of that alternative. Program managers need to understand all of the effects of each option so that a deliberate decision can be made with the appropriate transparency into the consequences.

As the examples demonstrate, understanding impacts is fundamental to properly managing a program so that the results of the analysis, actions, and decisions will match the expected outcomes. Without having understood these insights the results will be unexpected and the program team will be forced into a reactive mode all the time instead of focusing on the key activities required to meet commitments.

5.2.3 Authoritative Sources That Provide Transparency with a Customer Focus

Another common theme is the use of authoritative inventories of truth so that key program information is located and maintained in one place. This technique provides program stakeholders with transparency into accurate and easy-to-find information in a way that is relevant and understandable to them.

Figure 5.2 shows a view of the master inventories on a program. There are some inventories that should be at a program level such as the work intake pipeline and master project list. Other inventories such as financials and resources should be tracked at both a project level and a program level. Also note that there can be relationships between the inventories such as the one shown between resources and financials/schedule.

Each of the master inventories holds specific program information that needs to be maintained with current and accurate information. Table 5.1 lists these master inventories and what information is housed in them, essentially making that table the single source of truth for each of the single sources of truth.

By recognizing these sources of information as authoritative and centralizing them in a location where the team can access them easily, the team can all be on the same page regarding current program information and progress. Diligence and attention to detail in managing these inventories will then ensure that the information is accurate and relevant to keep them as living documents.

5.2.4 Tracking Progress for Early Indications of Problems

Another common theme is thorough tracking and monitoring of progress toward goals as a way to identify problems early (e.g., transparency). The intent is to identify trends so that the program team can have as much time as possible to take action and prevent the problem or just simply meet

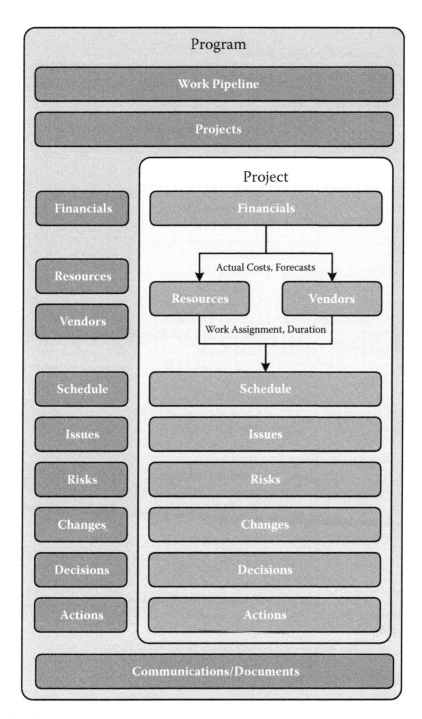

FIGURE 5.2
Single sources of truth.

TABLE 5.1

Single Sources of Program Information

Single Source of Truth	Information
Work Request Pipeline	• All program work requests with requestor information • Estimate history • Disposition and status of request
Master Project List	• All projects on program • Project manager name • Project description
Schedule (Project and Program)	• Deliverables and milestones • Key dependencies
Financials (Project and Program)	• Actual costs to date • Future forecasts • Variance analysis
Resources Roster (Project and Program)	• Resource names • Resource role • Organization
Resource Structure (Project and Program)	• Organization chart • Roles and responsibility matrix (e.g., RACI)
Vendor Inventory	• Vendor name • Contract information (e.g., contract number, start and end dates) • Invoice tracking
Changes (Project and Program)	• Change description • Impact of changes • Disposition and history for change
Risks (Project and Program)	• Risk description • Probability and impact • Action • Owners • Progress
Issues (Project and Program)	• Issue description • Impact • Actions to close issue • Owner • Progress toward closing issue
Decisions (Project and Program)	• Decision date • Description of decision • Stakeholders who made decision • Status and comments
Actions (Project and Program)	• Action description • Owner • Status of action • Progress toward completion

TABLE 5.1 (continued)

Single Sources of Program Information

Single Source of Truth	Information
Communications	• Stakeholder information
	• Communication methods
	• Frequency of communications
	• Owner
On-Boarding and Off-Boarding Checklist	• Key activities to bring on or roll off resources
Program Standards	• Processes on how to operate within program
	• Templates
Documents	• Program documents structure by project and phase

committed dates. Without effective monitoring, by the time problems are noticed there may be very little time to take action and then the program team will be in constant reaction mode. There are several examples of program functions that have areas to monitor closely:

- *Work intake.* Progress of the work request and estimate to meet commitment times or service level agreements.
- *Schedule.* Tracking progress toward meeting milestones and any early indicators of slippage against expected completion of activities.
- *Financials.* Monitoring overall program financial forecast as a function of actual costs incurred and future expected costs to understand trending against the original committed budget.
- *Changes.* Tracking the disposition and updating program documents based on the expected impacts of the change.
- *Risks.* Early identification and tracking of risks to determine actions to take to mitigate risk and then put them in place before the risk occurs.
- *Issues.* Recognition of risk and the management of the actions required to close it as quickly as possible to minimize the implications.
- *Decisions.* Tracking the decision and implementation of recommendations in a timely manner.
- *Action items.* Progress against the closure of open action items by the dates assigned to them.

There are many places for tracking and monitoring and the program team needs to recognize that these activities should be part of their roles on which they spend a significant amount of time. The implications of not tracking these items properly will be that the team will spend much more time on reacting to problems that arise and then explaining to senior management why they happened.

5.2.5 Mind-Set of Ownership, Leadership, and Stewardship

Program managers and team members must view their work with the mind-set of ownership, leadership, and stewardship. This approach will have them view their work in a different way and then also act differently as well. Programs are complex with many moving parts and therefore it becomes critical to have owners assigned to everything including milestones, action items, decisions, communications, changes, issues, and risks. These assigned people then need to be accountable for those items and drive them to closure without the need for someone else to follow up on them. Taking ownership means viewing dates as commitments that will be met and not just suggestions or targets.

Leadership means being a champion for the team and the goals of the program which can include facilitating difficult decisions, motivating the team members, providing guidance on the program, and influencing stakeholders through communications. Leadership does not just apply to the program manager as team members also have the ability to demonstrate these characteristics as well.

Stewardship is about caring for the program, the team members, and the company. There are many opportunities for this from just maintaining good documentation to helping team members grow their skills and find the right opportunities. This approach can also include providing the right information to make the best decisions for the company or just performing jobs well so that program commitments get met.

These critical mind-sets need to be recognized, expected, and also demonstrated by the program management team. Viewing one's self as accountable for work, a leader on the team, and a steward for the program and company changes one's perspective on the work and the approach to it.

5.2.6 Making Time for the Right Things

A program manager as well as the entire team must make sure that they prioritize their time to focus on high-value activities that will help them to meet program commitments. Throughout the book there have been many examples where the team has to be proactive and spend time on activities to ensure they get completed. Some of the important activities where a program manager needs to spend time include the following:

- Understanding trends of schedule and financial progress to ensure that the program is on track to meet commitments
- Looking ahead at the schedule to identify upcoming activities and then plan accordingly for them
- Ensuring that resources and vendors are available when needed with the appropriate skillsets
- Looking at health metrics to understand the progress of the project and key risk areas
- Proactively considering risks to the program and the proper mitigations
- Ensuring that issues, risks, changes, and actions are all tracking toward dates
- Using management by walking around (MBWA) to obtain critical information and build rapport with team members
- Communicating program information to stakeholders as well as facilitating decisions
- Preparing for meetings by gathering information, soliciting input, and considering the audience
- Ensuring that logistics and program operations are appropriate for program needs such as having available space to meet, a clear on-boarding process, and a standard location for documents

Program managers need to recognize that these activities are investments in time for the program and will prevent future problems and also make the entire team more productive. A program manager should consider the metaphor of a marathon where the team is running the race and the only job of the program manager is to stay ahead of the pack to make sure that the road stays clear so that they can keep running.

5.2.7 People Make the Principles Successful

Although the guiding principles are critical concepts to use on programs, their execution comes down to the people who are on the team and how they operate. It is not enough to say there will be diligence if there are not team members who operate with diligence. Program managers need to look for ways to improve the skills and mind-set of the team members to align with functioning under the guiding principles. There are several considerations and techniques that a program manager should employ to ensure that there is alignment between the concepts of the guiding principles and how team members operate.

- *Set clear expectations.* Program managers need to explain to the team that their expectation is that the team will operate by using all of the guiding principles. This is best done early in the program at a kickoff meeting or within the orientation materials to level-set the team from the start.
- *Align objectives.* Program managers should also look to align performance objectives with the guiding principles and their expected outcomes (e.g., transparency, diligence, and attention to detail).
- *Fit.* Sometimes team members are not in the right fit for their role and program managers need to consider how to optimally align resources to the work to build on their strengths. Where team members do not have the right skills, the program manager may want to consider training but note that some of the guiding principles are innate skills to how people operate and may be difficult to improve through training.
- *Hire people who demonstrate the skills.* Many times resources get hired because of knowledge of the business or of a specific technology and not based on how they work. Program managers should ensure that team members demonstrate the skills aligned to the principles and not just have the right business or technical background. A team member who knows a business process well but does not operate effectively can lead to a lack of productivity and quality that the program cannot afford to have.
- *Demonstrate the behaviors.* Program managers need to act in the same way they expect their team members to act. The approach will lose credibility if the program manager tells the team to pay attention to details and then does not do it himself.

The combination of operating under the guiding principles and having the right team members is really the key to success for any size program. Both of these require commitment from the entire program team and a lot of hard work but the result is a highly motivated and high-performing team.

5.3 TEMPLATE EXAMPLES

Throughout the book, there have been references made to standard templates and single sources of truth so this last section gives basic examples for some of the key templates. I use the authoring of my project management books as an example of a program throughout these exhibits.

5.3.1 Work Intake: Request Form

The intake request form is used during work intake to identify key information about the request which is then used to assess, prioritize, and plan the project. The example in Table 5.2 shows a request form for requesting that this book be considered as a new project in the program.

5.3.2 Work Intake: Pipeline Report

This report in Table 5.3 shows all of the projects in the work intake pipeline as well as where they are in the intake process and any key dispositions. The pipeline is the aggregation of the request forms plus additional assessment and disposition information. In this case it shows the two book publishing projects that I have managed.

TABLE 5.2

Work Intake Request Form

Request Item	Information
Request #	2.
Request Date	July 2012.
Requestor Contact Information	Kerry Wills, Author.
Description of Request	I am interested in a new project to write a book on program management guiding principles.
Value	The entire project management community will gather tremendous insights and have more successful outcomes.
Request	Estimate and commitment to start the project.
Request Response By	August 2012.
Commitments	Need to finish the book by February 2013 so it can be published and promoted during 2013.

TABLE 5.3

Work Intake Pipeline Report

#	Name	Type	Contact	Request	Due	Disposition	Estimate (hrs)
1	PM Skills Book	Commit	Kerry Wills	1/2009	1/2010	Approved	400
2	Guiding Principles Book	Estimate and Commit	Kerry Wills	7/2012	2/2013	Approved	500

TABLE 5.4

Work Intake Initiation Checklist

Check	Item
x	Add project to master project list.
x	Assign a project manager.
	Create the charter.
	Create the document storage location.
	Create the plan using the standard template.
	Update vendor master sheet with any new contracts.
	Assign resources to the project.
	Conduct kickoff meeting.
	Make formal announcement of project.

5.3.3 Work Intake: Initiation Checklist

Once a project is approved, a checklist is a helpful way to ensure that all initiation activities are known and tracked. The example in Table 5.4 lists several key initiation steps and shows that two of them are completed already.

5.3.4 Program Charter

The charter is used as the primary document for program scope, objectives, and assumptions. Most charters are written as text documents but the example in Table 5.5 has the information organized into key sections.

TABLE 5.5

Program Charter

Charter Item	Description
Program Description	• Write my second book on how to apply guiding principles to deliver programs effectively.
Business Case	• The benefits will be improved program delivery and more speaking opportunities for me.
Scope and Objectives	• Write a roughly 250-page book that includes case studies from real programs.
Assumptions	• The content is relevant to the audience. • The audience has a fundamental knowledge of project management techniques. • The book will have around eight case studies written by program managers.
Timeline and Work Plan	• Start Q3 2012. • Complete Q1 2013.
Organization	• Kerry Wills: Project Manager. • Kerry Wills: Author. • Multiple: Contributors. • John Wyzalek: Publisher.
PM Approach	• Document actions, risks, issues, and decisions in appropriate logs. • Communicate frequently with publisher. • Quality reviews with external peers.

5.3.5 Program Roadmap

A program roadmap organizes the major work for a program into milestones that may span several years. Table 5.6 shows key milestones and deliverables along a multiyear timeline for a program with two projects.

5.3.6 Master Project List

The master project list is the single source of truth for all project information on the program. This list, as shown in Table 5.7, can also include additional details on the projects such as financial information or milestones.

TABLE 5.6

Program Roadmap

Milestone	Project (Book)	Description
April 2009	Essential PM Skills	Start writing book.
January 2010	Essential PM Skills	Complete writing book.
July 2011	Guiding Principles	Create presentation on guiding principles.
May 2012	Guiding Principles	Confirm proposal for book.
October 2012	Guiding Principles	Sign contract and start writing.
February 2013	Guiding Principles	Complete writing the book.
July 2013	Guiding Principles	Publish book.

TABLE 5.7

Master Project List

#	Name	Description	PM	Cost (hrs)	Completion	Status
1	Essential Skills	First book on PM skills	K. Wills	200	January 2010	Completed
2	Guiding Principles	Focused on program delivery	K. Wills	300	February 2013	In Progress

5.3.7 Schedule

The schedule holds the key activity information including durations, resources, and dependencies. Table 5.8 depicts a project schedule with several activities and dependencies.

TABLE 5.8

Project Schedule

#	Activity	Start	Stop	Resources	Dependency	Status (%)
1	Write Book	10/1/12	2/1/13	K. Wills		100
2	Get Cases	12/1/12	2/1/13	Contributors		100
3	Create Presentation	1/1/13	3/1/13	K. Wills		100
4	Publish Book	2/1/13	7/1/13	J. Wyzalek	#1, #2	100
5	Promote Book	7/1/13	12/31/13	K. Wills	#4	50

5.3.8 Financial Tracking

The financial tracking model should be the single source of truth for managing project and program financials against budgets for specific categories such as resources and vendors. Table 5.9 represents one project but a program can roll these up.

5.3.9 Resource Roster

The resource roster shown in Table 5.10 is the master inventory of all resources on the program and also contains other pertinent information such as resource type, project assignment, or allocation.

TABLE 5.9

Financial Tracking Model

Cost Type	Budget To-Date	Actual To-Date	Variance To-Date	Budget Total	Forecast Total	Variance Total
People	$100	$80	$20	$200	$180	$20
Vendors	$200	$220	($20)	$400	$400	—
Materials	$50	$50	—	$100	$70	$30
Total	*$350*	*$350*	—	*$700*	*$650*	*$50*

TABLE 5.10

Resource Roster

Name	Role	Type	Project	Allocation (%)	Duration	Location
Brickhouse, Melissa	Contributor	Emp	Guiding Principles	10	11/2012–1/2013	Pennsylvania
Cordova, Amy	Contributor	Emp	Guiding Principles	10	11/2012–1/2013	Colorado
Wills, Kerry	Author	Emp	Guiding Principles	100	10/2012–2/2013	Connecticut
Wills, Randy	Contributor	Emp	Guiding Principles	10	11/2012–1/2013	Connecticut
Wyzalek, John	Publisher	Emp	Guiding Principles	5	10/2012–7/2013	New York

TABLE 5.11

Responsibility Matrix

Name	Write Book	Case Studies	Create Presentation	Publish Book	Promote Book
Brickhouse, Melissa	C[c]	R[a]		I[d]	
Cordova, Amy	C	R		I	
Wills, Kerry	A[b]/R	A	A	I	A/R
Wills, Randy	C	R	C	I	
Wyzalek, John	I	I	I	A/R	C/I

[a] R = responsible.
[b] A = accountable
[c] C = consulted.
[d] I = informed.

5.3.10 Responsibility Matrix

A responsibility matrix, shown in Table 5.11, is also known as a RACI diagram, and lists the key roles and responsibilities for a piece of work.

5.3.11 Vendor Inventory

The vendor inventory captures the key information regarding the vendors on the program and is the single source for all vendor and contract information. Table 5.12 shows two vendors who contributed to the project.

5.3.12 Change Request Form

The change request form captures all information regarding a proposed change control. This form is then used to assess and discuss the changes. Table 5.13 displays a proposed change to the project that has an impact on the schedule and resources.

TABLE 5.12

Vendor Inventory

Name	Work	Contact	Contract #	Dates	Status	Amount ($)
XYZ Consultant	Book graphics	Kontantin Nikolaev	123	10/2012– 2/2013	Open	1,000
ABC Consultant	Promotions	Thad Ozyck	456	1/2013– 9/2013	Open	500

TABLE 5.13

Change Request Form

Change Item	Description
Submitted Date	11/1/2012
Submitted By	Randy Wills
Request Description	Wants to add another section to the book which will be 20 pages
Project(s) Impacted	Guiding Principles Book
Impacts	• Schedule delay of two weeks
	• No financial impacts
	• Additional resources required for two weeks
	• Expected more benefits of satisfaction

TABLE 5.14

Change Log

#	Change	Impacts	Decision	Reasoning	Progress
1	Add another chapter	Schedule delay and resources for two weeks	Reject	We have enough content and this doesn't align with scope	Closed
2	Add templates to the end of the book	Additional schedule of two weeks	Approved	Not part of original scope but adds more value	Completed—changes made

5.3.13 Change Log

The change log is the inventory of all changes with information regarding the changes, impacts, and progress. The change request from above is added to the list shown in Table 5.14.

5.3.14 Change Impact Checklist

Once a change is approved there are several key deliverables that need to be updated so a checklist (Table 5.15) can be used to identify and track these items.

5.3.15 Risk Register

The risk register, shown in Table 5.16, is the master inventory to track all project risks and the actions taken against them.

TABLE 5.15

Change Impact Checklist

Check	Item
x	Update scope documents.
x	Update financial forecasts (if appropriate).
x	Update resource list (if appropriate).
N/A	Update vendor inventory (if appropriate).
x	Update schedule (if appropriate).
	Update benefits (if appropriate).
	Update status report.
	Update risk log.

TABLE 5.16

Risk Register

#	Date	Description	Impact	Owner	Due	Status	Actions
1	10/15/12	Case studies aren't done on time	Schedule delays	K. Wills	12/1/12	In Progress	Working with contributors to complete
2	11/1/12	Publisher doesn't like the structure or content	Schedule delays and rework	K. Wills	1/1/13	In Progress	Frequent reviews planned monthly
3	12/1/12	Book will not complete on time	Schedule delays in publishing	K. Wills	2/1/13	Open	Currently tracking to plan by 1/1/13

5.3.16 Issues Log

The issues list is the single source of truth for all project issues and for tracking the actions against closing them. The list is considered a living document that gets updated regularly. Table 5.17 shows one project.

5.3.17 Decision Log

The decision log captures all key decisions made and who made them on a project or program. The log, shown in Table 5.18, is helpful as an historical record for key decisions.

TABLE 5.17

Issues Log

#	Date	Description	Priority/Impact	Owner	Due	Status	Resolution
1	10/15/12	Graphics are in the wrong format	Medium - Schedule delays	K. Wills	11/1/12	Closed	Updated correctly
2	11/1/12	Contributor did not have case study done in time	Medium - Schedule delays	K. Wills	12/1/12	Closed	Found a different contributor

TABLE 5.18

Decision Log

Date	Project	Decision	Stakeholders	Assumptions/Rationale
10/1/12	Guiding Principles	Add case studies to book	K. Wills	Adds real-life credibility
11/1/12	Guiding Principles	Add template section to the end	K. Wills J. Wyzalek	More value to reader Demonstrates points from book

TABLE 5.19

Action Item Log

#	Date	Action	Owner	Due	Priority	Status	Comments
1	12/1/12	Add book to blog list.	K. Wills	2/1/13	Medium	Open	Will do once finished
2	1/1/13	Update link on LinkedIn profile.	K. Wills	7/1/13	Medium	Open	Once book is published
3	1/1/13	Update author profile on Amazon.	K. Wills	7/1/12	Medium	Open	Once book is published

5.3.18 Action Item Log

The action item log (Table 5.19) tracks all actions as they arise from meetings or other places and can be used to manage the closure of the actions.

5.3.19 Communications Plan

The communications plan (Table 5.20) stores all of the information regarding program communications.

TABLE 5.20

Communications Plan

Stakeholders	Information	Prepared By	Frequency	Format
Publisher	Progress against schedule	K. Wills	Monthly	E-mail
Contributors	Guidelines for case studies	K. Wills	As Needed	E-mail or call
Wife	Expectations on when I will leave my study for dinner	K. Wills	Daily	Yell out into hallway

5.3.20 On-Boarding Checklist

The checklist in Table 5.21 is used to bring on new program resources and get them up to speed as quickly as possible.

TABLE 5.21

On-Boarding Checklist

Check	Item
x	Add names to resource roster.
x	Add to e-mail distribution lists.
	Find a seat for the resource.
	Computer to do work.
	Access to key systems.
	Invited to key meetings.
	Announcement to team.

References

Akao, Yoji. 1994. *Development History of Quality Function Deployment*. Tokyo: Asian Productivity Organization.

Covey, Stephen. 1989. *The Seven Habits of Highly Effective People*. New York: Simon and Schuster.

Fleming, Q.W. and Koppelman, J.M. 2000. *Earned Value Project Management*. Newtown Square, PA: Project Management Institute.

French, J.R.P. and Raven, B. 1959. The bases of social power. In D. Cartwright (Ed.), *Studies in Social Power*. Ann Arbor: University of Michigan Press.

Grey, Clifford F. and Larson, Eric B. 2005. *Project Management. The Management Process*. New York: Irwin/McGraw-Hill.

Hubbard, Douglas. 2009. *The Failure of Risk Management: Why It's Broken and How to Fix It*. Hoboken, NJ: John Wiley & Sons.

Imai, Masaaki. 1986. *Kaizen: The Key to Japan's Competitive Success*. New York: McGraw-Hill/Irwin.

Keshishian, Mariette and Walkow, Patricia. 2010. *Where People and Projects Meet: Tools and Techniques for Understanding and Managing the People Side of Projects*. CreateSpace Independent Publishing Platform.

Nightingale, Jim. 2007. *Think Smart–Act Smart: Avoiding the Business Mistakes That Even Intelligent People Make*. Hoboken, NJ: Wiley.

Project Management Institute. 2011. *Practice Standard for Earned Value Management*, Second Edition. Author.

Project Management Solutions Inc. 2011. Strategies for Project Recovery. Glen Mills, PA.

de Saint-Exupéry, Antoine. 1939. *Wind, Sand and Stars*. Copyright renewed 1967, Lewis Galantière, (trans.). Orlando, FL: Harcourt.

The Standish Group International. 2011. *CHAOS Manifesto*. Boston, MA.

Index